Science and the Humanities

MOODY E. PRIOR

Science and
the Humanities

NORTHWESTERN UNIVERSITY PRESS
EVANSTON 1962

Open access edition funded by the National Endowment for the Humanities/Andrew W. Mellon Foundation Humanities Open Book Program.

Except where otherwise noted, this book is licensed under a Creative Commons Attribution-NonCommercial-NoDerivatives 4.0 International License. To view a copy of this license, visit http://creativecommons.org/licenses/by-nc-nd/4.0/

Moody E. Prior. Science and the Humanities. Evanston, Ill.: Northwestern University Press, 1962.

For permissions beyond the scope of this license, visit http://www.nupress.northwestern.edu/

ISBN 978-0-8101-3864-3 (paper)
ISBN 978-0-8101-3865-0 (ebook)

Printed in the United States of America

Original edition LCCN 62-13293

Acknowledgments

My first opportunity to try out some of the views which appear in these essays was in the fall of 1957 as a participant in a three-day symposium on science and the humanities held at Aspen by the Markle Scholars in Medicine. In the course of these discussions I came to appreciate how deep is the concern of intelligent men in this important issue of our times, and the exchange of views which the occasion provided brought out for me the peculiarly contemporary facets of this ancient question. To the Markle Foundation which made the occasion possible, to the Markle Scholars who invited me and provided an ideal audience, and to the men who took part in the discussions I no doubt am indebted for the origin of this small volume. In 1959, I was asked to give the faculty lecture at Northwestern University, and this opportunity provided the incentive to consider the matter seriously once more and to develop my views further. The problem of science and the humanities now became a preoccupation which I found difficult to set aside permanently for more pressing and immediate concerns. A leave of absence in 1960–61 granted me by Northwestern University, though devoted primarily to other matters, gave me at last the time necessary to get this book written. For his encouragement to present my views in book form I am indebted to Robert Armstrong of the Northwestern University Press.

Several persons have been good enough to read portions of this work in an earlier state, and their criticism and advice not only made possible numerous improvements in detail but also gave me the clues I needed to organize the discussion in its present form. For this invaluable service I wish to express my gratitude to Ronald Crane, Erich Heller, Dwight Macdonald, Robert Mayo, and J. Lyndon Shanley.

Moody E. Prior

Contents

Introduction

I have called the five essays which make up this volume *Science and the Humanities* in a deliberate effort to avoid implications of novelty. Often, the things which arouse our interest and occupy our thoughts do so not because they are new in essence but because they are immediately relevant, and they do not lose their hold on us, even though they may become tiresome, until they are resolved by common consent. The humanities and science in our time present an issue of this sort. The problem of the distinction between these two forms of knowledge and understanding and their relationship to one another is of long standing, but it is now again at the center of a vigorous discussion. Old issues, however, have a way of taking on new meanings with changes in the setting. During the late seventeenth and early eighteenth centuries the humanities assumed at times an aggressive critical role, and science found it necessary to reply in defense; today, it is the other way around, with science sure of its position and critical, and the humanities defensive and in rebuttal. Moreover, what was even thirty or forty years ago an interesting but comparatively relaxed controversy has now become urgent and intense and extends to national policy.

This group of essays does not undertake a systematic review of the current discussion of the humanities and science nor does it pretend to be an analysis in depth of the issues which the discussion raises. The essays are intended to serve a less ambitious and more immediate purpose. Since the last war, science has assumed a position of tremendous importance, and scientific activity has been enormously stepped up. The new position and new responsibilities of science have created many new problems, and these in turn have made necessary a good deal of soul-searching and public discussion on the part of scientists and of others who have found their interests involved or have a philosophic concern over the complex role of science in the modern world. Inevitably, the place of the humanities in these new developments has been touched upon, either as a central issue or as an incidental aspect of other matters. Some of the views and attitudes which have emerged are now becoming so familiar as to be taken for granted. Yet, under the stress of rapid developments and of the urgency of the problems which have arisen, not all statements have been thoroughly documented, not all assumptions have been critically analyzed, and—to put the extreme case—not all prejudices have been impartially examined. At this juncture, what seems to be called for is a reconsideration of some of the issues and assumptions of the current debate, and a revaluation of some of the conclusions toward which they lead. It is to this limited objective that these writings are addressed. I have therefore restricted myself largely to statements which have appeared within the last fifteen years, and in choosing them I have aimed at illustration of points of view that have emerged into prominence rather than at systematic documentation.

The five essays can be read independently of one another, but the order has been planned so that the ideas developed and the distinctions established in one section will provide support for later discussions. The first essay attempts to establish certain essential differences between science and the humanities and provides a foundation for the rest. The second considers the status of the humanities at the present time as a consequence of the rapid growth and extensive development of science and of the dominant position which it occupies today. One consequence of the present condition of the humanities is the criticism which has been directed against them by certain scientists and supporters of science. This phenomenon is examined in the third essay. The fourth and fifth essays are concerned with two practical issues which have come into prominence within very recent years—the national importance of science and of the scientist in present day affairs, and the scientific education of the non-scientist. These sections may at first glance seem detached from the first three, but the relevance of the more abstract discussions which precede the last two essays should, I hope, be apparent even though I have not pointedly insisted upon it.

One of the disturbing aspects of the current intellectual milieu is the strong partisan spirit which seems to animate so much of the discussion about science and the humanities. It is as though one is forced to write for the humanities and against science, or in the interest of science and against the humanities. The prevalence of this contentious spirit has, I believe, had something to do with the confusion and lack of philosophic clarity which occasionally mars the prevailing discussion of the question. Though I cannot help writing from the point of view of one who is not a scientist, I have certainly not written as an opponent of science. The direction of the present inquiry, however, has been unavoidably influenced by the fact that, because of the high and strategic importance of science, it is the scientists who have taken the lead in the recent commentaries on the role of the humanities in relation to science. For the most part, therefore, it is the views of scientists and their supporters that I have had to examine critically, and the state of the question being what it is, I have at times unfortunately had to proceed argumentatively. If I can judge from the sensitiveness which I have detected in a few of the recent writings in the cause of science, I will be extremely fortunate to escape the charge that I have a bias against science. For the few partisan spirits who may respond in this way, I wish to protest in advance that I find the products of scientific and technological activity fascinating, that I appreciate the comforts and refinements of living provided by this activity, and that I enjoy the company of scientists and engineers, many of whom I admire and respect. I have no bias against science as such, and I do not believe it possible or expedient to repudiate modern science and retreat to some primitive utopia, like that, let us say, of eighteenth-century Europe, a comparatively simple age still living in the happy memory of such recently departed culture heroes as Copernicus, Galileo, and Newton. I have written in the confidence that for the most part scientists will certainly understand the imperative necessity of subjecting everything to critical inquiry, if only as a means of sharpening the terms of discussion. Scientists cannot help being impressed with the importance of their work and the fabulous character of their activities, and it is their obligation to seek the advancement of science; but enough of them have expressed a responsible concern over the place of science in our world to indicate that they would not wish, any more than others, to encourage an idolatry which makes "the service greater than the god."

Science and the Humanities

I
Science and the Humanities: An Essay in Definition

In the current discussions of science and the humanities, definition of the two key terms is not often undertaken. For the most part, it is tacitly assumed that such distinctions as are necessary to any given discussion can be taken for granted as part of the reader's general qualifications for following the argument, and that in any case they will emerge by implication. This approach is understandable. Definition is a tedious matter, and the refining of distinctions belongs to Polonius. Dictionary definitions are next to useless, and anything more serious requires not only linguistic and philosophical distinctions, but, often, reference to developments in cultural and intellectual history. During the course of any extended debate, however, distinctions become blurred, and under the stress of argument issues are shifted and nuances disappear. It becomes necessary now and again to reconsider what it is we are talking about.

What is science? What are the humanities? What distinguishes these two forms of creativity, these two ways of bringing order and meaning to the data of experience? What are the methods peculiar to these disciplines, the boundaries by which each is circumscribed, and the special qualities and powers characteristic of each? The way in which these questions are formulated is important. It is science and the humanities which are the subject of inquiry and not, for the purpose of definition, scientists and humanists. Often, this separation is not maintained, and a great deal of confusion arises, some of it simply annoying, some of it serious. A few of the common ways in which this confusion occurs need to be considered in order to eliminate this source of obscurity.

One of the topics which encourages the shift of subject from science to scientists is the relation of values to humanistic and scientific learning, an issue currently much discussed. Commenting on this question, James Bryant Conant writes:

> A conscious effort on the part of many investigators to control disease, to prolong life, and to alleviate pain has yielded results of a most dramatic nature. I should like to point out that the conduct of almost every individual who participated in this advance of science and progress in the art of healing was determined by a set of value judgments. These judgments were closely connected with the exhortation, "One ought to help the suffering." The conduct of doctors, we all know, is regulated by a set of ethical principles which in themselves are based on value judgments. What I am here emphasizing is the degree to which a judgment of value has determined the investigation of scientists or those seeking by empirical means to improve an art. Once again I make the point that those who say that science and value judgments are in separate compartments have failed to examine the nature of scientific undertakings and the motivation of many scientists.[1]

The proof that science and human values are necessarily related cannot be found in the conduct of scientists who have advanced knowledge or in the motives of those who have participated in the progress of scientific learning. The shift from science to scientists has taken place in this statement without any recognition of the fact that such a transference of subject has altered the nature of the problem. The altruistic motive—"one ought to help the suffering"—is not inherent in the nature of science, though it may be a property of scientists; indeed, the most that Conant can say in proving his point is that it is "the motivation of *many* scientists." It may equally well be the motive of many statesmen, poets, or literary critics, though again it may not. The motives of individuals in the pursuit of their profession vary greatly. The humanist as well as the scientist may wish to advance knowledge out of a desire to improve the lot of mankind, but both may also be inspired by the pleasure to be derived from the successful solution of a problem and the exercise of disinterested intellectual effort, or merely by the desire to refute an unpleasant rival. These are variables, depending on the vagaries of human nature and of the individual temperament. Medical investigators, the group selected by Conant to make his point, may conceivably be more often inspired by a desire to alleviate human suffering than theoretical physicists. The latter may exert themselves rather in the hope of becoming Nobel laureates, but as far as the possible benefits to mankind are concerned their motives may prove to be quite as valuable in the long run. Unless we wish to end up defining science as what it is that scientists do, and the humanities as what it is that humanists do, we will have to keep separate the human drives which impel any given human effort from the special characteristics of the kinds of products which result from those efforts.

In trying to make the proper distinctions between science and the humanities, it is also necessary to suspend consideration of the effect of the pursuit of these disciplines on character or conduct. The point is frequently made, in refutation of a particular view of the nature of the humanities, that the qualities attributed to the humanities are not necessarily found in those who devote themselves to humanistic learning: humanists are sometimes narrow, pedantic, and uncultivated; and on the other hand scientists are often cultivated, humane, and sensitive. "It is impossible," remarks David Riesman in a characteristic comment, "to tell whether a man is a humanist or not by the label of his disciplines; and I have seen a number of colleges where anthropologists are more humanistic than the teacher of English, the physicists far more humanistic than the economists and sociologists." [2] There is nothing in the statement which any observer of the world of learning would care to dispute, whatever his discipline. Moreover, in some contexts this datum has its importance. And to the extent that the excessive claims of teachers of humanities—especially those of an older generation—for the value of their discipline have made this form of rebuttal necessary, reiteration of the point may be said to have its place. The damage occurs when such assertions intrude into the effort to determine the nature of these two presumably distinct forms of human creativeness. It is one thing to point out that a particular form of learning possesses certain latent powers, and another to point out that those who are exposed to that form of learning do not always possess the qualities implied by those powers. There is never a perfect equivalence between the potential formative influence of a particular discipline and the individuals which have come under its influence. Moreover, the inferences hidden in such statements are ambiguous, and the rhetoric serves to give them a force which the writer may not have intended. The trouble with most such statements, whether about science or about the humanities, is that they convey implications about the character of these disciplines while presumably only making off-hand remarks about the vagaries of human behavior. In any effort to define in an impartial spirit the distinctive characteristics of science and the humanities, such assertions amount to a diversionary tactic which is capable of introducing an element of incoherence in the analysis by an unnoticed shift of the subject from science to scientists, humanities to humanists.

In a similar way, the problem of making necessary distinctions is further complicated by the preoccupation with the relationship of science to culture.[3] As it appears in recent discussions, the question has several aspects—whether science is to be regarded as an important aspect of present day culture, whether the study and pursuit of science are of equal merit and value with the arts and letters in cultivating the sensibility and imagination of the individual, and whether scientific men do or do not possess the qualities of mind and temperament associated with the idea of a cultivated man. These various aspects of the matter often shade into one another. Jacob Bronowski, who has written with knowledge and insight on the place of science in the modern world, complains that, in his efforts to make clear the importance of scientific study to the well-being of society and the development of the modern individual, he is often stopped

> by those whose education and tastes are literary, because they find these claims puzzling. They know what culture is: it is Sophocles and Chaucer and Michelangelo and Mozart and the other figures at the base of the Albert Memorial. And they know what culture is not: it is not laundry lists and sleeping pills and the proved reserves of oil and the *Statistical Digest*. In short, culture is not a body of facts: but what is science but a body of facts? How then, they ask, can science be a part of culture? There is no scientist at the base of the Albert Memorial.[4]

Similarly, Ernest Nagel observes, "scientific inquiry is frequently believed to be a routine grubbing for facts, and unlike literature and the arts to require no powers of creative imagination."[5] One is tempted to inquire, by whom? I myself can think of no non-scientist in whose company I would wish to spend any time who thinks of science as a collection of facts, who believes that science that is any good requires no powers of creative imagination, and who would exclude science from any idea of culture considered as the aggregate of the main artistic and intellectual activities of a society or an age. The argument along those lines is, one would hope, out of date. Matthew Arnold found it necessary to explain that, in defining culture as the best that had been thought and said, he assumed the inclusion of such scientific achievements as those of Newton. Today there would not be much dissent from the following statement in *Science and the Creative Spirit*, a book written by a group of non-scientists:

> What science has done is significant in its own right, in comparison with the fine arts, with the long calendar of the world's literature, with the attainments of scholarly humanism, and with the humanist preoccupations of the philosopher. We believe therefore that the true humanism of today demands that science be included in the full account of mankind, that its contribution to man's life in freedom, to the creative and non-utilitarian part of man, be set down in order.[6]

In part, the concern over this matter on the part of the scientists arises from the use of the word, "culture," denoting the arts and letters as chief among those activities and accomplishments of a society or an age which are not related to its survival or material well-being and which represent the exercise of the mind and spirit without reference to the material utility of its products. This use of the term has been for many years associated almost exclusively with the arts and letters because these were the activities which conspicuously contributed to the non-material accomplishments of society, and persons were described as "cultured" because they had sufficient leisure or incentive to develop an interest in these products of their culture. There are obvious historical reasons why this restricted use of the word "culture" did not embrace participation in science or familiarity with science at a serious level, and in some contexts the word still tends to exclude professional interest in science. In its limited popular sense, however, the word is beginning to lose caste, and persons who have a serious interest in arts and letters tend to shy from "cultural" as a characterization of their activities and "culture" as their product. In some popular contexts the words suggest refined leisure, but not the serious activity of serious minds. I suspect, moreover, that many literary persons would regard absence from the base of the Albert Memorial as a matter for self-congratulation. To some scientists, this implied exclusion of their craft from the accomplishments of "culture" has become translated into a belief that as a class they are not regarded as cultured or cultivated persons. "The scientist," writes Eugene Rabinowitch, "is often represented as a narrow specialist with little understanding and interest in literature and art, in history and social problems, and utterly naïve and gullible in his political convictions. . . . It is easy to point out that this picture is inapplicable to many scientists—in fact, to a large majority of them; that many—particularly the really outstanding ones—are very widely read; that the most abstract of them—the theoretical physicists—are, as a rule, highly musical." [7] A non-scientist probably can have no real appreciation of the extent to which this feeling on the part of scientists is justified; my own observations lead me to believe that scientists are needlessly over-sensitive on this score. The important consideration at present, however, is that this preoccupation with science and culture conveys certain implications about science and the humanities while dealing primarily with scientists and humanists, a context which inhibits a neutral approach to the nature and powers of the two disciplines.

Where this source of confusion does not arise, the chief difficulty lies in the variety of definitions currently in use and the diversity of approaches to the problem of definition. On the surface there seems to be no agreement concerning the special subject matters and methods which may be said to be characteristic of science and the humanities. However, a review of definitions currently in use, while it reveals some irreconcilable differences among the various approaches to definition, does bring out certain common elements among them which can serve as the starting point for an effort to establish accurate and useful areas of meaning around the two basic terms.

It is a common practice today to divide the empire of learning into the sciences, the social sciences, and the humanities—a division which influences the administration of universities and the organization of college catalogues. In the division of the humanities are included languages, literature, art, music, and usually history and philosophy (the occasional association of these last two with the social sciences has some small philosophical basis but is usually the result of administrative convenience in colleges and universities). These are in large part the subjects which, until the great flowering of science, formed the basis of a "liberal education." Since these divisions are the result of convenience and tradition, they are taken for granted without any very precise or explicit justification. The division appears to follow distinctions relating to subject matter and content, but since it has grown up in colleges and universities it also implies some relationship to educational goals and hence further implies something about the distinctive methods and inherent powers of these various disciplines. Emphasis on one or the other of these aspects of the humanities—whether methods or powers—will affect the kind of definition which emerges in consequence. If the emphasis is on the means employed, the humanities can be thought of as characterized by certain intellectual methods which may be applied to any subject matter in order to reveal its distinctive meaning and its relevance to human experience and use. This approach underlies the old division of the academic curriculum into the grammar, rhetoric, and logic of the trivium. From this point of view, science may also be included among the humanities, if the concern is not with mastering the operational methods of science and its present content, but is rather with the philosophic understanding of its methods, its relationship to other branches of learning, and its influence on the thought and culture of the past and of our times—as an aspect, that is, of epistemology, esthetics, history, and the like. The emphasis can also be placed on the ends rather than the means, in which case the humanities can be regarded as those forms of art and learning which are capable of developing certain human powers in producing a desirable human ideal—an ideal variously conceived in the course of history as, for example, Plato's philosopher, the Renaissance courtier, or the citizen of a free society. Today, such objectives are likely to be stated more broadly, as in the following statement of the ends of humanistic study by Howard Mumford Jones: "The humanities are those branches of knowledge (and activity) that have a special capacity, if rightly interpreted by humane learning, to mature the intellectual and moral powers and to quicken the sensibilities of the individual." [8]

Some confusion arises from the fact that the humanities are seen to include the arts as well as certain forms of scholarship, activities presumably different in method and aim. In the arts, the claims of imagination and insight are validated without having to be justified by scientific criteria. The same cannot be said for any form of scholarship. To the extent that certain kinds of scholarship are devoted to an understanding of the arts, they may be readily accepted as humanistic by association. The humanities, however, also claim certain independent forms of scholarship that call for the evaluation of evidence, critical and even skeptical rigor of thought and analysis, and rational ordering of the results —for instance, history, philosophy, or such works of comment and information as the observations of de Toqueville on America and those of Burke on the French Revolution. Such scholarly writings are sometimes thought of as constituting a distinct category, a bridge between the humanities and sciences.[9] For certain purposes, it may be convenient to consider them in this way, and it has the advantage of reducing the danger of oversimplification and of calling attention to the great variety of synthetic and interpretative forms available to the human mind. They are not essentially scientific in spite of their scholarly character, and they display properties which are found in works which are recognized as distinctively humanistic.

When we search for the common element among the various types of work usually regarded as humanistic, as well as among the various approaches to a definition of the term, the humanities will be seen to include those forms of art or learning which are directly concerned with human responses to all forms of experience, and therefore primarily with those aspects of human experience "that cannot be resolved into either natural processes common to men and animals or into impersonal forces affecting all members of a given society." [10] Inseparably involved in them are questions of human uses and goals, of the ends toward which men direct themselves and the means they use to gain them. Accordingly, they are drawn to such issues as beauty and ugliness, happiness and misery, right and wrong, good and evil, and the like. They reveal the potentialities of men as human beings, and reflect on the possibility of the full realization on these through feeling, thought, and action.

There is some diversity also in the approaches to the definition of science. It is sometimes maintained that the mark of true science is exact measurement, a point insisted upon by Kelvin. Unquestionably, the most exciting aspects of modern science would be inconceivable without the increasingly exact and subtle capacity to measure, but there are certain aspects of science—for example, classical biology and geology—which are excluded if measurement becomes the basis of differentiating science from other forms of activity. Recently, there has been a disposition to emphasize as a distinctive feature of science the fact that it is cumulative and characterized by continual progress, features which it is claimed the humanities do not possess. As a recent popular expression of this notion has it, "Today's writers deal with questions of right and wrong, beauty and ugliness, human nature and the universals of life, while Greek playwrights of over 2,000 years ago wrote on exactly the same themes." [11] Fundamental concepts are, of course, timeless in any discipline: today's scientists are still studying motion, time, and space, as were the ancient Greeks. There is, moreover, a sense in which the non-sciences are also cumulative. A non-science like music affords the musician of today vastly greater and more complicated resources than were at the disposal of a composer of medieval tropes. Moreover, certain achievements in the arts are seminal and productive of further developments, as are significant scientific theories and concepts. There is a further form of cumulative artistic development illustrated by a work like Joyce's *Ulysses*, which is fully comprehensible only with reference to Homer's epic and the place of *The Odyssey* in world literature. Science is indeed cumulative and progressive, but this aspect does not provide by itself an unmistakable basis for differentiation from certain non-sciences. [12] In its crude form this view of science is not very useful, but we can see some of its possibilities as a clue to the meaning of science in Conant's notion of science as progressive, as a "guide to action":

> In view of the revolution in physics, anyone who now asserts that science is an exploration of the universe must be prepared to shoulder a heavy burden of proof. To my mind, the analogy between the map maker and the scientist is false. A scientific theory is not even the first approximation to a map; it is not a creed; it is a policy—an economical and fruitful guide to action by scientific investigators. [13]

> Science is a dynamic undertaking directed toward lowering the degree of empiricism involved in solving problems; or, if you prefer, science is a process of fabricating a web of interconnected concepts and conceptual schemes arising from experiments and fruitful of further experiments and observations. [14]

Underlying this conception of science are an appreciation of the revolutionary and dynamic character of recent science, and a preference for the esthetic justification of scientific discovery rather than the utilitarian or "engineering" justification. The common view of science as an exploration of the universe is rejected, and is replaced by the notion of science as "a guide to action." Science is conceived of as simply a form of intellectual activity which prepares for further activity of the same sort. Nevertheless, Conant's statement allows some room for inference as to what kind of action and what kind of guide.

We may take as a starting point a statement by the philosopher, Ernest Nagel, of the distinctive intellectual goals of science: "It has been the perennial aim of theoretical science to make the world intelligible by disclosing fixed patterns of regularity and orders of dependence in events." [15] The formulations of science have been deemed successful to the extent that they incorporate all phenomena of a particular category and predict the course of future events of a similar order. This aspect of scientific achievement is suggested in Jacob Bronowski's description of science as "the organization of our knowledge in such a way that it commands more of the hidden potential in nature," [16] as well as in Conant's phrase, "lowering the degree of empiricism in solving problems." Science accordingly aspires to the highest possible order of probability, and sometimes the probability is so high as to amount to virtual certainty. Scientists protest with reason that there is no such thing as scientific certainty, and point in evidence to the overthrow of previously uncontested scientific "truths," instances in which a particular scientific formulation was confronted with a category of events which it could not subtend, and a new organization of data had to be sought. The present emphasis on the progressive and dynamic aspects of science and the accompanying denial of certainty have obscured the fact, however, that the rejection of rigorously demonstrated older scientific theories is something less than total. It is, in fact, one of the merits of scientific method and of scientific proof that any formulation which has fully met the tests science recognizes as requisite for the validity of a proper generalization never quite loses its claim to scientific truth. It is commonplace today to say that modern science has repudiated Newton, as though

Newtonian physics were now in the same limbo as the Ptolemaic theory. Modern science has had to set aside Newtonian physics as inapplicable and irrelevant to the phenomena and the areas of investigation which are the modern scientist's present concern, but within a certain category of events Newton is still valid, as the artificial satellites attest. All science approaches universality, certainty, and predictability as its goal, though it is at the same time recognized that there may never be an end to the road leading there.

Before exploring some of the implications of these approaches to definition, it is necessary to call attention to characteristics which both science and the humanities have in common. As manifestations of the creative activity of man, they both attempt to reduce to intelligible relations particulars of experience and observation according to some principle of order and selection. Moreover, although scientific method is a rigorous formalized instrument for controlling observation, verifying results, and directing the kind of synthesis which may be achieved, the creative act is not essentially different in science from that in the arts. In the most original products there is the same brilliant grasping of the points of similarity and relatedness between different observations and experiences, as well as the use of fruitful analogies—the experience which is summed up in the word "intuition." Bronowski finds a formula for science in Coleridge's definition of beauty: "Science," he writes, "is nothing else than the search to discover unity in the wild variety of nature—or more exactly, in the variety of our experience. Poetry, painting, the arts, are the same search, in Coleridge's phrase, for unity in variety. Each in its own way looks for likenesses under the variety of human experience." [17] In spite of the close association between practical technology and science, the scientist and the artist are both impelled by the pleasure of creativeness: "For most scientists," writes Conant, "I think the justification of their work is to be found in the pure joy of creativeness; the spirit which moves them is closely akin to the imaginative vision which inspires an artist." [18] And there is also an esthetic element in science which is experienced not only in the apprehension of the tremendous synthetic power of its laws and formulas but also in the tracing through of the steps of an "elegant" experiment or demonstration. The creative act, comments Bronowski, "is the same in Leonardo, in Keats, and in Einstein. And the spectator who is moved by the finished work of art or the scientific theory relives the same discovery; his appreciation also is a recreation." [19] Comparison of the products of scientific and humanistic activity reveals some surprising interrelations. There are certain respects in which some forms of science resemble the arts, in particular music, and in those respects science has more in common with the fine arts than does humanistic scholarship. At the same time, humanistic scholarship resembles science in one important respect; both are circumscribed by the demands of accurate observation and verification. From this point of view humanistic scholarship has more in common with science than with the fine arts with which it is normally associated among the humanities.

There are those for whom these interesting similarities between science and the humanities suggest the ultimate unity of all forms of knowledge and all forms of creativity, with the result that distinctions between science and the humanities become meaningless and the differences are rendered inconsequential. That they are not meaningless is indicated by the persistent retention of the terms even by those who labor to obliterate the distinction between them. Science and the humanities do not ask questions of the same kind, and the generalizing powers of the humanities tend in a somewhat different direction from those of science. In science, the total absorption of the individual event in the generalization is the goal; on the other hand, the humanities are concerned rather with providing for the special meaning of the individual event within an appropriate general system. This is the case with such diverse products as poems, plays, and histories. The special merits of the generalizing powers of science are responsible, moreover, for the fact that the final fruits of science are impersonal and transferable to other purposes and other contexts in a way that humanistic achievements are not. It is true, as Bronowski points out, that if we take the whole of a great scientist's work it bears the unmistakable mark of his peculiar genius, in its successes as well as in its failures; and it may therefore be possible to say, from this perspective, that "science at last respects the scientist more than his theories, for by its nature, it must approve the search above the discovery and the thinking (and with it the thinker) above the thought." [20] But when we consider the final product which has received the formal and technical approval of science, it appears impersonal, independent of its creator, and can be adopted as the property of any scientist who needs to use it. Granting the existence of a discernible regularity in the phenomena being studied, it is conceivable, too, that some day the gravitational formula would have been discovered even if it had escaped Newton. None of this can be said of a sonata of Beethoven's or a play of Shakespeare's, or for that matter of a history by Thucydides. These possess a uniqueness not characteristic of the discoveries of science. And there is this further distinction between the ultimate products of scientific and humanistic creativity, that the formulations of science are necessarily indifferent to the question of their human use or meaning; they convey no direct or implicit comment on the goals which men may choose or the means which they may employ to attain them.

This last consideration is at the center of one of the most sensitive issues in the current discussions of the nature of science and its place in the modern world. It arouses intense responses in defense of the human relevance of scientific discovery, and the place of values in science and the scientific activity. One aspect of this argument has already been alluded to—the view that human values inhere in science since many scientists are inspired in their researches through a desire to relieve human suffering. It is a position as old as Bacon, who pronounced charity as the prime virtue of his scientist because he viewed science as an instrument for gaining control over nature to the end that the lot of man might be improved.[21] This view became outmoded as the emphasis shifted to the ideal of the scientist as a man engaged in the disinterested search for scientific truth, but it has recovered some of its appeal today, even for those who do not give a primary place to practical and humanitarian applications in their definition of science.[22] The weakness of this line of argument, as we have already indicated, is that it shifts the subject from science to the scientist and thus alters the nature of the problem from the character of the product to the character of the individuals who create it. But the objection to the idea of science as impersonal and neutral does not rest solely on the notion of the possible altruism of the scientist's motives. Science demands accuracy and eschews error—it is concerned with truth. It demands honesty and the subordination of individual wishes and prejudices. Values are certainly involved here. Science, moreover, has been viewed not as a merely personal but as a collective activity that has created a world-wide community governed by mutual respect for its members and encouraging originality and freedom. These virtues did not, of course, originate with science and are not unique with scientific activity, and they are evident in a conspicuous way as a concomitant of scientific activity because they are a necessary factor in its success. As Bronowski remarks in an extended discussion of this issue in *Science and Human Values,* "They have grown out of the practice of science because they are an inescapable condition of its practice." [23] The final products of scientific activity, however, are by their nature dissociated from the values and personal virtues which were involved in the human activity which produced them. In one sense, of course, something of the same sort can be said for a humanistic activity, such as literature: the submission to the discipline of writing, the respect for good craftsmanship, the desire for honesty, even at times the endurance of privation which may be demanded for the creation of artistically successful novels, plays, or poems are not necessarily involved in a proper description of what literature is. But the significant difference can be seen in the fact that the final product of scientific activity is impersonal and uncommitted in any way to any particular human use or goal; the final product of literary effort, on the other hand, is inevitably identified with its author's character and his personal artistry, and it cannot escape its involvement with particular human feelings and with a particular view of human conduct and human aspirations and goals.

There are certain differences between the methods and aims of science and the character of its distinctive products compared to those of the humanities, and it is important to insist upon them because the tendency in much of the present day discussion of these matters has been to obliterate or obscure these differences. Among these differences, one of the most conspicuous is that, unlike the sciences, the humanities are concerned with emotional responses to experience and they evoke these responses; and this is true of all the arts and in varying degrees of most humanistic writing. Another difference is that the humanities address themselves to an understanding and an evaluation of human goals, and this, while less apparent with the most abstract arts like music, is especially apparent in all forms of humanistic discourse. By comparison, a scientific generalization, whether a mathematical formulation, or a theory, or a concept, carries no implication within itself of its relevance to any human uses to which it may be put, to the human choices which may be governed by it, or to the inherent human striving for happiness or self-fulfilment in action— except as it points to further scientific activity. The creations of science—its mathematical syntheses, its proven generalizations, its fruitful concepts and theories—are neutral with reference to their moral and social implications; but with the humanities, the involvement in both the human meaning and response to the experiences and observations dealt with is inescapable and is inherent in all typical humanistic products. They could not, in fact, be described or defined without reference to these.[24] Science and the humanities share in common the capacity to arouse a particular form of esthetic response—the pleasure which is induced by those products of creative effort of whatever kind in which discrete elements of matter or experience are brought together in a meaningful organization and which delight with the sense of difficulty overcome. Even in connection with this aspect, however, there is a difference: it is possible to define a product of scientific activity without reference to this esthetic aspect, but it is hardly possible to do so for a work of music or a poem.

There is a further difference. Though both attempt to discern and express order and unity in the variety of experience, their generalizing powers tend in different directions. Science attempts to subdue a multitude of incidents to a grand generalization which, until challenged by new events that demand a different order, is universal in its application and is capable of accurate prediction with reference to all future events that belong to the category of incidents with which it deals. Humanistic works, on the contrary, are concerned rather with the individual experience, and they relate it to general principles not in order to have it lose its identity within them but in order to reveal its special meaning. Humanistic works therefore tend toward uniqueness, and in their totality call attention to the diversity and plentitude of human experience. All sunsets can mean only one thing in any given statement about them in the science of physics; to a painter each sunset is a distinct phenomenon to be given a special personal meaning in the whole range of sunsets, and is subject to the widest range of artistic orderings, and capable of many emotional evocations. There is also a parallel difference in the resultant products of the two forms of creative activity, that whereas the products of scientific genius are in their final form impersonal, the products of artistic genius are unique and inseparable from the special powers of the one mind which produced them.[25]

Some of these essential differences become evident when we compare the humanistic and scientific approaches to human activity. The effect of a scientific ordering is to produce detachment from the individual experiences which are being dealt with; the effect of a humanistic ordering is to encourage involvement.

A simple illustration may help to give concreteness to this distinction. The National Safety Council of America has for some years engaged in the fascinating business of predicting the number of traffic casualties which will occur during a given national holiday. One of its most ambitious productions was its prediction in 1951 that the one millionth traffic fatality since the first recorded automobile death in 1899 would occur in December. This prediction was first announced in March. By December 1, the Council had zeroed on the target; the millionth fatality would occur on

December 22. The purpose of the prediction was to call attention to the relentless progress of this slaughter and to encourage caution and consideration. If such admonitions had had any immediate effect, the consequence would have been that the authority of the methods of prediction would have suffered, but as it turned out the statisticians had nothing to fear. On the appointed day the destined number of wretches gave up their lives on street and highway that the prophecy might be fulfilled. But for some reason, the event did not produce a significant reaction. In spite of an elaborate campaign, the final event received surprisingly scant notice, and such editorializing as appeared was admonitory in a conventional way. The only genuine response which I encountered came from an individual who said, "Well, thank God, that's over. Now we can all breathe more easily again." Save for this original touch, the whole affair was empty of serious human meaning. The preliminary publicity was aimed at leading up to a solemn response to an awesome national tragedy. The public response was not as to a tragedy; it was the kind proper to a scientific generalization. The lifeless body on the highway was a mere number representing the operation of inexorable impersonal laws and paying tribute to our genius for accurate measurement. One could not expect the response to have been otherwise. The closer an intellectual synthesis approaches the scientific ideal, the more completely will the human act lose its individual significance, and the more fully will we remain detached from its human meaning within the formulation of which it has become a part. Where numerical formulations are involved, this effect becomes especially noticeable.

The contrast to this may be seen in literature, for the power of literature lies in its capacity to involve us in its data in a predetermined way, as well as in its capacity never to lose sight of the uniqueness of the individual experience which informs the work. There is a respect in which literature shares with science the capacity to formulate concepts which give us a new outlook and provide a new measure of control over our observations. The world takes on a different aspect after mass and gravity and evolution have been conceptualized, and in an analogous fashion we apprehend experience differently once *Hamlet* or *War and Peace* has been conceptualized. It accordingly becomes possible to apply even the concepts provided by literature to the organizing of varieties of experience. Hamlet, for instance, has been frequently used as a prototype of a particular form of modern sensibility. But a modern writer who sees Hamlet as a symbol of dislocated and troubled man in modern society is employing an illuminating analogy, and this is not quite the same as a scientist fitting a physical event into a general law. Outside the limited analogy, *Hamlet* remains unique. And within the play itself, the universals to which Hamlet's actions are referred do not function to obliterate the distinctiveness of his tragedy. On the contrary, his own uniqueness gives a special meaning to these universals. Unlike the creations of science, which are of necessity neutral with respect to their human meaning or use, works of literature of necessity involve us in such responses as pity, fear, sorrow, pleasant and bitter choice. We can remain neither detached from nor indifferent to their human meaning. In successful works of literature, our involvement is so complete that they attach our sympathy even where they do not compel our intellectual conviction or belief. Antigone's compulsion to bury her brother, if only with a handful of earth, has its origin in a world of taboos alien to our own, yet her tragedy moves us, at this remove from classical Athens. Because they can do this, works of literature have the capacity to extend the range of our sympathies; they impress upon us the diversity of human experience and direct attention to the values which determine individual choice and through which human actions acquire their meaning.

It does not—to state the obvious—follow that a man of letters or a student of literature and the arts and of humanistic learning is of necessity more humane and wise and perceptive than one who is not a humanist, any more than it follows that a scientist, by virtue of his practice of science, always thinks more logically, clearly, and impartially than the non-scientist. It is reasonable to suppose that continuous involvement in the discipline of science will leave its impress on the way a man thinks, and that the demands which the practice of science make upon him will shape his character.

By the same token, it is equally reasonable to suppose that the arts and humanistic learning will contribute their share to shaping the attitudes of those who take a serious interest in them. And both, imperceptibly, leave their mark on the age in which they flourish and on the society which gives them support and scope for their activities. At best, however, we can only infer the direction which the influence of these two creative approaches to experience will take, through an understanding of the essential character of their products and of the ways in which they differ from one another. And perhaps we can express only negatively the possible consequences of their respective powers with any assurance or accuracy. There are important functions which the humanities cannot perform, and there are important functions which science cannot perform, if for no other reason than that they do not ask the same kinds of questions. The humanities cannot take over all the methodological procedures of the sciences nor duplicate the comprehensive inclusiveness of scientific generalizations and their capacity for accurate if limited prediction, and we cannot, therefore, expect that the humanities will provide exact and fully operational solutions to the problems that vex our human condition. Science lacks the capacity of the arts, especially literature, and of humanistic learning, to become preoccupied with proper human goals and proper human means of attaining them, or to create a concern for the individual experience and to search for its human meaning; and we cannot, therefore, expect its distinctive contribution to lie in the direction of keeping alive and encouraging a sense of our common humanity.

II
Science and the Present Status of the Humanities

The prevailing attitude toward the humanities may be likened to the mixture of concern and contempt provoked by a once distinguished relative who is now old and poor and therefore something of a nuisance and a moral embarrassment to his more fortunate kin. An interesting history could be written of the origins and development of the critical attitude toward the humanities, and of their decline, if not fall, in our century; but for the present purpose no elaborate documentation is called for. All informed persons today are aware of a chronic concern for the humanities, and suspect with varying degrees of conviction that the tremendous importance of science has something to do with the present status of the humanities. The suspicion is, of course, not unfounded. The growth of modern science has been one of the most impressive phenomena of modern times, and some would regard it as the most important single development of the last three centuries. The arts and humanistic learning have always been responsive to their intellectual and social environments. The subject matter which they use, the problems they deal with, the new form which they give to recurrent universal questions, are all affected by the political and social conditions and by the intellectual climate of the times. All these factors have been greatly influenced by modern science, and in that way modern science has, directly and indirectly, become involved in the character of the arts and of humanistic learning today. But there is an additional important consideration in the fact that these developments have had the effect of making an important issue of the character and the relevance of the humanities themselves. The present state of the humanities is in large part a consequence of the vigorous interaction between a dynamic and rapidly developing science and all other branches of learning and thought. Precisely how this has taken place has not yet been fully documented; but it is possible to place certain important developments in relief by viewing the effects of present day scientific activity against those of the sixteenth and seventeenth centuries—the great age of Copernicus, Keppler, Galileo, Newton, a period much like ours in the appearance of men of genius, in the incidence of impressive discoveries of originality and imaginative brilliance, and in the revolutionary character of its achievements.

The first scientific revolution of modern times was characterized by an extraordinary impetus to study nature, which resulted in discoveries in many fields—for instance, Harvey's discovery of the circulation of the blood. But the achievements which establish the special character of the science of this age were in astronomy and physics. Another of its distinctive features was that, although accident and even wrong presumptions played a part in the advance of this science, there was a good deal of self-conscious philosophic concern for method, illustrated in such works as Bacon's *Novum Organum*, Descartes' *Discourse on Method*, Galileo's *Dialogue Concerning Two New Sciences*, and Boyle's *Sceptical Chymist*. In consequence, investigation of nature was directed by excellent theorizing on the nature of science, the limits within which it could properly operate, and the means by which the discovery of scientific knowledge could be promoted. The activity of science was limited to the searching out of the behavior of matter—to the world of "extension," to "secondary causes"; science could not be appropriately applied to questions relating to God and the soul, to questions of faith, and to metaphysical and moral issues depending on these. This delimitation of the field was reflected in the distinctive methodological features of the "new science," as it came to be called. These features were the insistence on exactness and accuracy of observation aided by instruments to overcome the limitations of the senses, the use of experiments, the formulation and testing of hypotheses which appeared to "save the phenomena," and the application of mathematics to the study of motion. The model which guided the method and shaped the results of this science was atomistic and mechanical.

The splendid success of these new approaches to the study of the physical world is to be found in the by now familiar record of the scientific discoveries of this age. What were some of the consequences of these advances in knowledge? One very important consequence was that the science of the sixteenth and seventeenth centuries radically altered the picture of the universal setting of men's lives.[1] The universe being revealed by the new science was endowed with a new vastness which appears to have made a profound impression, illustrated, for instance, in the sober wonder with which Robert Boyle sets down the figures which had been computed for the distance to the sun, moon, and planets.[2] The cosmos had, however, become not only larger but more impersonal in consequence of the impression that the essential character of the universe was its mathematical and mechanical perfection. "God geometrizes" is a phrase that appears frequently, and a common analogy for the cosmos was the clock. Established cosmologies become the nucleus for a variety of attitudes and beliefs, so that any radical change in the picture of the universe appears to be a direct attack on the attitudes and beliefs themselves. The new cosmos of sixteenth- and seventeenth-century science seemed to threaten old reassuring feelings and convictions which rested on a different view of the universe and which appeared to have been deprived of their foundations. Even before the picture of the new universe was complete, a vague sense of unease had appeared, reflected in the familiar line of John Donne, "new philosophy calls all in doubt." By the late years of the seventeenth century, the universe appeared to be so mechanically perfect as to require no resident engineer, and so vast as to inspire a sense of cosmic loneliness. It was feared that the new science would encourage atheism on the model of the ancient pagan atomists, and in Thomas

Hobbes, who followed mechanistic principles consistently in all directions, pious men saw an enemy who demonstrated how their fears might indeed be realized.

The questions and apprehensions raised by these aspects of the new science were resolutely attacked by some of the scientists themselves, as well as by others who accepted the new science and its methodology but who wished at the same time to preserve some of the reassurances about man and his world that had accumulated around the old order. They pointed out that the new scientific explanations were for the most part hypotheses to account for phenomena and not metaphysical certainties, and they further cast doubts on a purely mechanistic view of the universe by calling attention to phenomena for which mechanistic theories were unable to account. At the same time, attacking the problem from the opposite direction, they insisted that the regularity and mechanical perfection which were being revealed in nature were the surest proof of a rational order in the universe and, therefore, of God. If the seventeenth and early eighteenth centuries were spared the kind of rough warfare of science and religion which racked the nineteenth century, it was partly because the scientists were eager to support the theologians and the theologians were receptive to resolutions of a rationalistic order.

The new science also had an influence on the conception of man's intellectual capacities and future possibilities. The older view of the limits of human knowledge was colored by traditional pessimism from two sources. One of these was the religious caution that man should not aspire to know too much, since such aspiration stems from pride and is improper to mortals. The other was classical skepticism, which rejected the possibility of accurate or adequate knowledge because man's capacities are defective and his life short. The skeptical critique proved more serious than the religious, and discussions of scientific method usually reviewed the skeptical position with respect and proposed ways of overcoming the sensory and intellectual defects of man. In a bold step, skepticism was adopted as a method while being rejected as a philosophical system, a position which Bacon expressed with his usual felicity in *The Advancement of Learning:* "If a man will begin with certainties, he shall end in doubts; but if he will be content to begin with doubts, he shall end in certainties." The scientific view also eliminated the sense of discouragement in the idea that life is too short to permit adequate knowledge by viewing knowledge as cumulative and its growth and approach to certainty as continuous beyond the limits of any single lifetime—the fulness of knowledge, as Bacon put it, lay in the fulness of time. This incipient idea of progress also supported the view that the increase of knowledge about the physical world, when properly applied, would bring about genuine improvement in man's lot. Bacon spoke of regaining the lost Paradise. For the first time there could be seriously entertained the possibility that man need not always remain in physical want and discomfort. This aspect of scientific thought suggested a progressive view of knowledge and encouraged optimism concerning man's capacities and man's future.

As the methods and attitudes of science rapidly gained in authority, they were borrowed in other branches of learning. The mechanistic model, and to some extent the methods of relating phenomena to it, are apparent in the psychological theories of Hobbes and Locke. Thinking in traditionally humanistic fields was colored by the presumption that there might be some grand unifying principle, like Newton's discovery about gravitation, which would bring order to knowledge in areas outside the physical sciences. There was, of course, nothing new in the conception of a rational and immutable order in nature operative not only in the physical world but also in law, morality, art, and the forms of society. In fact, this ancient notion was one of the important inheritances from humanistic thought of the past that was taken over as a guiding assumption by science itself. Nevertheless, the splendor of the idea in its new guise was irresistible and there were borrowings—some of them naïve—of scientific presumptions in other fields, contributing a new accent or coloring to the general search for laws and principles "according to nature." This analogizing is illustrated by occasional references to the ethical principle of benevolence as "moral gravitation." These developments were not, however, very far-reaching, for scientific methods and attitudes did not fundamentally alter humanistic modes of thought nor appreciably affect the status of the humanities. There is apparent only a suggestion here of what was at a later date to become an influence of extraordinary significance.

Somewhat more indirectly, the humanities were affected through the influence of science, and the speculations it aroused, upon the sensibilities of the times. Milton retained the Ptolemaic system of astronomy in *Paradise Lost*, but his feeling for space is Copernican. The nature poetry of James Thomson is permeated with the images and philosophic overtones introduced by science, as in the following passages from "A Poem to the Sacred Memory of Isaac Newton" (1727):

> Even now the setting sun and shifting clouds
> Seen, Greenwich, from thy lovely heights, declare
> How just, how beauteous, thy refractive law.
>
> What wonder than that his [Newton's] devotion swell'd
> Responsive to his knowledge! for could he,
> Whose piercing mental eye diffusive saw
> The finish'd University of things,
> In all its order, magnitude, and parts,
> Forbear incessant to adore that Power
> Who fills, sustains, and actuates the whole.

In Addison's Ode beginning, "The spacious firmament on high" can be seen the happy resolution of the disquieting elements in the new cosmology. The last stanza concludes:

> What though, in solemn Silence, all
> Move round the dark terrestrial Ball?
> What tho' nor real Voice nor Sound
> Amid their radiant Orbs be found?
> In Reason's Ear they all rejoice,
> And utter forth a glorious voice,
> Forever singing as they shine,
> "The Hand that made us is Divine."

The poem is little more than an eighteenth-century version of the psalmist's "The heavens declare the glory of God," but Addison infuses into the old images the new problems and the new resolution. The sense of terror in the new universe is urbanely minimized in the subordinate construction of the first part of the last stanza, and the ancient notion of the harmony of the spheres is made to serve the purposes of the argument from design in its modern scientific form.

Science further affected the humanities by contributing to the forces which were shaping prose style in the direction of clarity, simplicity, and directness. The interest of the scientific writers in prose style was not primarily esthetic, but was dictated by their preoccupation with method and their concern for the reformation of learning. Their criticism of medieval learning was that it was disputatious and verbal, and they aimed to separate true learning from argument and from persuasion through appeal to the emotions, and to turn attention back to things and not words, as the cliché of the times had it. They urged the avoidance of literary ambiguity and encouraged exactness, and they aimed to write, as Sprat put it, with a mathematical plainness. The influence of this aspect of seventeenth-century science on prose is now generally recognized. Equally important, however, was the fact that the new scientists introduced a renewed interest in language as a symbolic tool, and gave to this ancient problem a direction which has led eventually to the subtleties and refinements of modern linguistic and logical study.

Many accounts of this first period of modern science stress its struggles against opposition and misunderstanding. There is the wretched episode of Galileo with the church; there is the long reluctance to accept the Copernican system; there is the effort of the Royal Society to promote the acceptance of science and justify its value. There was opposition to science—more vigorous in England than in France—that took the form of philosophic objection to its aims and methods and satiric ridicule of its claims and pretentions. Some of this was directed at the amateurishness and zeal of the minor devotees of experimental science, but much of it stemmed from a conviction of the superior value to man of humanistic learning against the scientists' claims for the certainty and value of scientific knowledge. This view persisted for some time; Swift could still, early in the eighteenth century, write truculently about science and even with some disparagement about Newton. But Swift's friend Pope was closer to the mood of the day when he summed up the almost universal feeling about science and Newton in a famous epigram:

> Nature and Nature's laws lay hid in night.
> God said, "Let Newton be," and all was light.

Opposition there was, but what stands out in historical perspective is how readily, in fact, the field was won. The impact of the new concepts and speculations aroused by science was almost immediate, and their diffusion outside the circle of those who had a direct interest in science was, everything considered, quite rapid. Especially after Newton, it is easy to trace the positive effects of a triumphant new science on the thought and on the sensibility of the age.

Comparison of this account of the first great revolution in science with similar developments in our own age has, in addition to the limitations of an oversimplified view, the disadvantage of neglecting significant episodes which occurred during the intervening years. The basis of comparison is therefore impoverished, especially by the neglect of the radical new theories in biology and geology and their influence on religion and ethics and on the conception of the social sciences and history. In compensation, however, there is the advantage of a sharpening of relief by concentrating on the extremes of the historical continuum.

We are struck at once by a parallel. Although the areas of activity encompassed by the physical sciences in our century have increased and appear to be without bounds, with extraordinary discoveries being made in all areas, the most exciting fundamental developments have been, as in the earlier period, in astronomy and physics. And in Einstein our age also has its hero, who sums up for us the distinctive powers and accomplishments of our science as Newton did for his own day. Like the science of the earlier period, moreover, that of our age has been revolutionary. It has swept aside a great deal that was formerly accepted as true about the physical world, and has brought about a radical revision of the picture of the physical universe. These parallels are striking, but they can be misleading if pushed too far. The innovations of recent science constitute a revision and not, as in the earlier age, a repudiation of preceding science. They reflect an increase in the quantity and exactness of observation and in the subtlety and sophistication of scientific thinking, but they are in effect a continuation of the revolution which began in the sixteenth and seventeenth centuries. They are regarded as a reaffirmation of the rightness of scientific activity and of the vitality of its scientific principles and traditions, and hence, paradoxically, the rejection of the relevance of much earlier science has increased confidence in science itself. The image of science has been accommodated to this role. Emphasis is placed on the dynamic and progressive aspects of science, the idea of scientific certainty has been modified, and it is assumed that the picture of nature will be continually and radically revised.

The changes which have been introduced involve a rejection of earlier models and the introduction of more subtle and complex hypotheses. The clock will no longer serve even as a crude analogy. The nature of the modern scientific universe, in comparison with that of Galileo and Newton, has become elusive, even whimsical. Newton, commenting on God and nature, wrote in a letter to Bentley in 1692, "to compare and adjust all these things together in so great a variety of bodies, argues that cause to be not blind or fortuitous, but very well skilled in mechanics and geometry." Contrast this with a remark of Einstein's: "Raffiniert ist der Herr Gott, aber boshaft ist er nicht"—of which one translation is, "God's tricky, but he ain't mean." The science of Galileo and Newton made a strong appeal to common sense; that of today sometimes seems to defy it. Because the technical foundations of modern science are complex and recondite and because its methods and generalizations sometimes seem at odds with the logic of everyday practical experience, it is not so readily accessible to educated non-scientists as was that of the seventeenth century. This is a condition which has created serious concern and apprehension among scientists and non-scientists alike. In the years immediately following Newton, it might have been supposed that a person well educated by the standards of his day could comprehend the most advanced science available to him if he were willing to devote the proper study to it. This was surely the assumption of Henry Pemberton, editor of the third edition of Newton's *Principia*, in his book, A *View of Isaac Newton's Philosophy* (1728), a systematic presentation which attempts to give a comprehensive account of Newton's contributions to science, and which does not spare the reader the necessary, if simplified, mathematics, and whose list of subscribers is a *Who's Who* of the age. In contrast, efforts to explain the contribution of Einstein and his contemporaries to non-scientists have to resort to parables and metaphors, and it is not uncommon for physicists to insist, with a feeling of resignation, that beyond a certain point it is impossible to understand what modern science is really about without being a scientist. On the one hand, the triumphs of modern science have created great confidence in science and interest in its discoveries, and on the other, they have by their nature created a barrier between the scientist and his non-scientific fellows.

Certain developments in modern science have, it is true, stimulated new speculations in areas of thought traditionally humanistic and religious in a way reminiscent of similar developments in the past. By comparison, however, the new efforts seem pallid, and the over-all result of taking over scientific concepts and theories for these purposes has been disappointing. The concepts of relativity, indeterminacy, and frames of reference have become a part of popular learned vocabulary, but for the most part they function as little more than figures of speech and lose their original exactness and creative power in the transfer. A similar lack of vitality characterizes the speculations which have been prompted by the new image of the universe in support of theological and moral generalizations. The concept of indeterminacy has been appealed to in arguing for the freedom of the will, and the absence of complete certainty and predictability in the case of a certain order of physical events has—in a paradoxical reversal of the seventeenth-century pattern—been advanced as offering new evidence for a belief in God. None of these efforts, however, has been comparable in its influence to those of the earlier age, when the argument from design and a few casual remarks by Newton about God and space made a profound impression on theologians and educated laymen.

There are reasons, aside from the difficulty and obscurity of modern science, that account for the lack of vitality in the transference of scientific concepts and discoveries to humanistic and religious areas. There are, after all, some impressive aspects of the modern scientific view of nature that can be appreciated by men who have some imagination even though they have no understanding of the means by which this picture of the universe was arrived at. One reason for the difference in our response to new scientific discoveries and that of the earlier age is that the shock of novelty in the changing picture of nature is not so great as it once was. We have come to associate our idea of science with the recurrence of astonishing revelations about the physical world, and we are now apparently conditioned to accept these without metaphysical fears so long as we are assured that they are based on experimental evidence and mathematical proof. It may be symptomatic of this attitude that the traditional warfare of science and religion is quiescent, and that the major religions of the Western world have entered the twentieth century in harmonious association with science and eager to demonstrate their modernity in this respect. They have shown greater nervousness over the new psychology than over any of the remarkable discoveries of physics and astronomy, or even of biology. The advances of science create a fearful response only when they threaten established assurances about man and his position in nature. The philosophic adaptations and reassurances which followed the scientific discoveries of the seventeenth, and then the nineteenth, centuries have come down to us as familiar concepts, and *mutatis mutandis*, they continue to be serviceable. The awe of Boyle over the magnitude of the planetary system is not of a different order from that of a modern man scanning the photographic star surveys taken with our big telescopes and wondering over the light years that are required to measure astronomical distance; for although the scale of the cosmos has been magnified both in the infinitely small and the infinitely great, no one appears to be terrified as was Pascal by the silence of those infinite spaces, bemused on the contrary by prospects of cosmic exploration. The world revealed by modern science has astonished the imagination, and it is as yet too early to appreciate fully the extent of its influence, but one thing seems reasonably sure—we have not found it necessary to orient ourselves all over again in order to be reconciled to our cosmic habitation.

These observations do not imply that new discoveries of such scope as those of modern science have had little or no influence on man's thoughts about himself, the effects perhaps impressing themselves in ways not yet clearly apparent or fully realized. They are merely a warning that it is necessary to distinguish the trivial from the important influences, the presumed from the real. It has been frequently suggested, for instance, that modern science has confronted man with such overwhelming demonstrations of his puny insignificance as to render humility imperative. "It is not easy," writes Bertrand Russell, "to maintain a belief in one's cosmic importance in view of such overwhelming statistics." [3] Perhaps it is not easy, but most people manage to do it somehow. They continue to behave as though their concerns are important in the universe. Such humility as the moralist of science seeks to instill today is not new. "When I consider thy heavens, the work of thy fingers, the moon and the stars, which thou hast ordained; What is man, that thou art mindful of him?" The cosmos of modern science has at most added a new coloring to this theme. Sensitive and imaginative men throughout the centuries have found humility in the contemplation of the magnitudes and mysteries of the universe, but such men have always been few and they are few now in spite of the supergalaxies. The effects of modern science on man's conception of himself are to be found elsewhere, and one of these is in the direction away from humility. The discoveries of modern science have been so stimulating and our breath-taking scientific progress so pregnant with the expectation of future discoveries, that the principal response has been a feeling of enthusiastic admiration for man's capacity to lure nature's secrets from her. Moreover, the association of science and technology is no longer accidental, as it largely was in the past, but systematic and purposeful, and the products of this union have been so numerous and successful as to bring about an unprecedented advance in man's efforts to extend his empire over the physical world. And there seems no end to the discoveries and their application. In no other enterprise has man shown himself so lordly, and in no other have his efforts reflected such credit on his intellectual powers and resulted in so many tangible consequences which testify to his abilities. Confidence in scientific endeavor is reflected in the support which it now receives, and the tempo of scientific research is being stepped up with something approaching frenzy in the effort. Optimism concerning man's powers and the possibilities of scientific study is apparent today on a scale which dwarfs that of the early pioneers who introduced it.

This is very probably the only area today where optimism exists. For at the same time, the very success of the scientific effort has bred a contradictory feeling of uncertainty concerning man's capacity to make proper use of the great powers which have become available through science. These fears have been expressed for some time, largely because of an apparent lag between the discoveries of science and society's capacity to put them to use without creating problems for which answers could not readily be found. With the invention of nuclear weapons, however, this fear has become a very real terror of the existence of powers which could destroy civilization if not the race of man. The Faustian myth has once again assumed a tragic aspect.

The general effect of the progress of modern science has not been to engender humility, but rather to promote at the same time a curious contradiction of optimism and fear— optimism concerning the unlimited capacity of man to uncover the secrets of the universe and to extend his dominion over nature, and fear and apprehension over the tremendous powers which man's ranging intellect has placed in his hands.

These apprehensions are indirectly associated with science in another way. While the success of science has increased confidence in itself, it has at the same time undermined confidence in other forms of intellectual activity, and especially those which concern themselves with human problems and with man's philosophical needs.

It is the humanities which traditionally have addressed themselves to the formulation and, hopefully, the resolution of questions and apprehensions involving the human condition, and it is some indication of their position today that they have been criticized for failing to provide answers to the dilemmas which confront modern man, in particular those stemming from the success of science. Ironically, the humanities have also been complained of because they have failed to adopt the methods which have proved so fruitful and successful in science. Such criticism fails to appreciate the real position of the humanities in respect to science today. Many commentators express disappointment because they fail to discover a simple direct relationship between particular innovations in science and some philosophically interesting or useful notion in humanistic thinking or their translation into some ingenious motif in the arts. The most far-reaching effect of science upon the humanities is to be sought in the triumphant progress of science itself—in the high authority of scientific methods and the penetration of scientism into every remote corner of our intellectual activity, in the identification of reality with the models useful to science, and in a preference for the kinds of problems which lend themselves to study by means of the methods and instruments employed by science.

The natural sciences have become synonymous today with accuracy, exactness, certainty, trustworthiness, and true knowledge. Whatever can claim kinship, however remote, with science assumes these virtues and partakes, if only by a dim reflection, of the prestige which the whole scientific syndrome commands today. The widespread diffusion of this attitude is reflected in the unsophisticated acceptance of the symbols of science as a kind of popular folklore. Whereas the confidence of our grandparents in patent medicines was inspired by the reassuring faces of the Brothers Smith, Lydia Pinkham, and Father John, today confidence in proprietary drugs is created by the picture of an alert looking individual in a lab coat performing a laboratory rite, or by a naïve diagram or animation and the meaning-less repetition of formulae and technical words. I am per-suaded that no one is convinced at the rational level by any of this and that its effectiveness is largely ritualistic, celebrat-ing the imago of science and thus acknowledging its great power and authority. It does reveal, however, that at a low popular level, science inspires belief and confidence com-bined with a touch of awe.

At the intellectually respectable level, science has exerted a profound influence upon our conception of what consti-tutes adequate and acceptable knowledge. Speaking of the presumptions which governed his work in the 1920's and 1940's, Bertrand Russell tells us that among these was a "prejudice in favour of explanations in terms of physics wher-ever possible." [4] The interesting word in this casual remark is "prejudice," for it denotes not so much a settled convic-tion as an ingrained preference, and one which would be widely and sympathetically understood. One consequence of a prejudice in favor of explanations in terms of physics is a prejudice against questions which cannot be answered in those terms. The world of learning today generally tends to be mistrustful of problems which do not lend themselves to investigation by scientifically approved means. Yet the issues which interest the humanist cannot for the most part be reduced to elements that can be investigated scientifically. An interesting sidelight on the position of the humanities is provided by the current state of academic philosophy. The traditional philosophic questions have for some time ceased to excite much interest, and the most flourishing kind of philosophy at the moment is represented by the various schools of analysis, for which the primary problems of philosophy are linguistic. According to the most rigorous ex-ponents of one school of modern philosophers, the kinds of propositions normally involved in humanistic discourse turn out upon investigation to be merely rhetorical or emotive utterances which cannot become the basis of any contribu-tions to knowledge because they defy verification.

An illustration of one way in which this point of view af-fects the humanities is provided by the case of I. A. Richards. As co-author of *The Meaning of Meaning,* he is in full sympathy with the modern scientific approach to knowledge. He is also a student of poetry, which he evidently likes and approves of. He faces this dilemma in *Science and Poetry.* Poetry turns out to be valuable in putting in order the vagaries of our primordial psychological and emotional mechanisms. This provides a useful role for poetry, but it is nevertheless a comedown for the poets who not too long ago were declared the unacknowledged legislators of mankind. Today they are the acknowledged non-legislators. Meaning has become scientific meaning, knowledge, scientific knowledge. The primacy of scientific learning affects poetry, and in fact most of the arts, in still another way. Scientific knowledge is concerned less with "what" than with "how." It explores a reality which finds its ideal expression in mathematical relations, and cares for the individual experience only as it becomes a clue to the possibility of an impersonal formulation in which the individual instance loses its identity in an all-powerful generalization. In those arts which are concerned with the concreteness of individual objects and events and with the distinctiveness of the individual unique experience, this change in emphasis concerning what constitutes significant reality and true knowledge leaves the artist with a feeling of being outmoded, and with a sense of remoteness from the intellectual center of his times. This feeling has been expressed indirectly and directly by numerous poets since the early nineteenth century, and it appears to have driven some of them to an almost obsessive inwardness of theme and expression.[5] It has also encouraged a view of the arts as no more than pleasant sophisticated diversions.

In those areas of learning which are concerned with man and his activities, now almost universally referred to as the social sciences, the principal characteristic today is the adoption of scientific methods and models. Whereas in the seventeenth and eighteenth centuries the influence of the methodology of science in humanistic learning was suggestive and tentative merely, today it is widespread. The social sciences have separated themselves as far as possible from the humanistic associations of their past, and have taken over the techniques and attitudes of the physical sciences and applied them to human and social phenomena of all sorts. They construct models and devise experimental situations. They have become quantitative. They aspire to scientifically verifiable generalizations sufficiently certain to permit accurate prediction, and their failure to achieve generalizations comparable in significance and dependability to those of the natural sciences they attribute to their youth as sciences. And in keeping with their scientific status, they avoid implications of value in their formulations.[6] In this they are certainly right, for to the extent that they are scientific and hope to achieve scientifically respectable results, they cannot allow themselves to determine, in consequence of their purely scientific efforts, the proper human uses of their discoveries. The success with which the natural sciences are able to isolate particular problems and solve them has encouraged the view that, if the social sciences with their present scientific orientation could keep pace with the physical sciences, they could solve and thus eliminate the problems created by science and technology, as well as such evils as war or the unequal distribution of resources.

This is an understandable but misguided hope. *Scientific knowledge* would be greatly advanced if the social sciences could achieve the accuracy and certainty and the power of fruitful generalization characteristic of the physical sciences, and like other advances in knowledge this would be a positive gain whose consequences cannot now be predicted. But if the present hopes for the social sciences are realized, the consequences of research in the social sciences will confront us with exactly the same kinds of problems as now face us in consequence of research in the natural sciences. Like the discovery of nuclear fission, they could eventuate in good or evil, depending on whether we were wise enough to put them to use in such a way as not to destroy whatever is decently human in us. The discoveries cannot determine their own use. The benefit of increases in scientific knowledge is that they provide a clearer view of the circumstances within which action is possible, and thus they both limit and enlarge the possibilities for real choice, but they do not provide within themselves any guide for the kinds of choices to be made, since such choices raise the question of human goals and the means which are proper to attain them. Issues like these have been the perennial preoccupation of the humanist, but his questions are not scientific questions, and he cannot therefore employ all the distinctive methods or be guided by the aims of science without repudiating his role and nullifying what he is after.

The dominant efforts to resolve this difficulty acknowledge, either directly or by implication, the success of the scientific view of knowledge. There are three principal approaches which seem to have attracted attention to themselves recently, none of them free of difficulties. One of these places the ultimate source of human values and the meaning of experience in religion. This, probably the oldest approach, has survived centuries and has vital personal and philosophical advantages. Difficulties arise because religions tend to become authoritative and mutually exclusive. In a pluralistic society or world, therefore, there must be found certain common principles based on acceptable reasonable sanctions in order that men can act in concord. A second approach distinguishes two separate orders of truth, one based on scientific method and another, different from science and equally valid, which is discovered not through scientific method but through myth and metaphor. The approach is intuitional and anti-rationalistic. Intuition is an avenue of discovery in all fields—"The symbol and metaphor," writes Bronowski, "are as necessary to science as to poetry" [7]—but intuitions require to be demonstrated within a rational system or framework before their truth can be generally acknowledged. Not all intuitions have proved verifiable, and not all myths which have strong emotional appeal have proved to be good or valid. In any case, the mythic approach to knowledge, in spite of interesting philosophical support, has not made great headway against the prevailing scientific attitudes.

A third approach has come through science itself. Attempts to demonstrate that scientific method can be applied to the determination of basic human values have proved ineffectual, but it has been proposed that these values can be derived from the activities of scientific investigation. Bacon had observed that the pursuit of science calls for the highest virtues, and his views were echoed during the seventeenth century. Recently it has been suggested that the human qualities demanded by science of its practitioners can become the foundation for a consistent body of ethical imperatives and norms:

> The society of scientists is simple because it has a directing purpose: to explore the truth. Nevertheless, it has to solve the problem of every society, which is to find a compromise between man and men. It must encourage the single scientist to be independent, and the body of scientists to be tolerant. From these basic conditions, which form the prime values, there follows, step by step, the spectrum of values: dissent, freedom of thought and speech, justice, honor, human dignity and self-respect.[8]

The international community of scientists, in its candor, mutual respect, and absence of social and racial discrimination, constitutes an ideal democracy: "Though there is an aristocracy of talent and unequal distribution of powers and prestige among scientists, the organization of science as a community of free, tolerant, yet alertly critical inquirers embodies in a remarkable measure the ideals of a liberal civilization." [9] Presumably, in scientific activity we possess an empirically demonstrated and realistic basis for a code of conduct and a social order which does not require religious or metaphysical sanctions and which has proved its validity through the success of the scientific activity and its influence on the conduct of scientists as a community. A vital consideration here, however, is that this code governs scientific activity only because it has a special relevance to science as science. As Bronowski puts it, the scientific virtues "have grown out of the practice of science because they are the inescapable conditions of its practice." [10] That is to say, they are as necessary to a scientist as co-ordination is to a dancer and the acceptance of danger is to an explorer or mountain climber. Upon what sanction and with what justification or authority can we insist upon their general adoption in areas of human activity where they are not the inescapable conditions for successful practice—in politics, for example, or the competition for markets, war, love? We have today evidence that where a society provides the scientist with the kind of freedom and respect which he requires for success within the restricted area of his special activity as a scientist, it may deny them to the society of which he is a part, including the scientist himself as member of this society, without any perceptible effect on the progress of science or the ability or willingness of the individual to exercise his full powers as a scientist. C. P. Snow doubts whether the conditions are as vital even to the society of scientists as the older generation of scientists maintained.[11] Assuming, however, Bronowski to be correct in his view that "the society of scientists must be a democracy," it does not therefore follow that such privileges as "free inquiry, free thought, free speech, tolerance" will transfer themselves to society as a whole, by example or necessity, in a scientific age. The virtues called for by the scientific activity are impressive and it would no doubt be for the general good if they could be transferred to other activities, for they were admired and recommended long before modern science; but it is only another indication of the authority which science commands to suppose that the exercise of these virtues could be generally adopted on no other authority.

When we pass from what science is, and what it has contributed, to the implications of these activities and accomplishments for human choice and action and for the human condition, there is involved a leap from one order of intellectual systematizing to another. In the present state of the world of learning, it sometimes appears as though there is no choice between knowledge that comes as a command of God or an unverifiable intuition, and science. Not all rigorous, critical, disciplined thinking, however, is science, for what characterizes science is not only the distinctiveness of its methods and the rigor of its thinking but also the kinds of questions which are being asked. Failure to perceive this distinction leads either to the superficial adoption of the paraphernalia of physical science as a masquerade where they do not function essentially, or to the discredit of serious, critical, impartial inquiry where the questions are not scientific and the results cannot claim the same degree of probability and predictability as is possible in science. Yet the possibility and necessity of such thinking need to be acknowledged, and on this matter it is useful to have support from a scientist, James Conant:

> Only an occasional brave man will be found nowadays to claim that the so-called scientific method is applicable to the solution of almost all the problems of daily life in the modern world. Yet some proponents of this doctrine have at times gone even further and maintained that only by a widespread application of the scientific method to the problems of society at every level can we hope for peace and sanity.[12]

However, the point of view expressed here is not universally held and its implications are not generally understood. What we might call the collective consciousness of our times continues to have a prejudice in favor of explanations in terms of science. In addition, according to the most rigorous analytical philosophies of our times, not merely are the traditional methods of humanistic learning not entirely respectable, but the questions which are its concern lack exactness and are usually meaningless because they are beyond proper verification. Analytical philosophy has thus provided support for the undermining effect of scientism on the validity of humanistic approaches to experience. While the more extreme aspects of this view are being challenged today, the destructive implications of this analysis have not been repudiated in such a way as to eliminate the ambiguity of status which attaches to humanistic learning in consequence of the acknowledged supremacy of science and its methods and the widespread adoption of scientific attitudes.

In perspective, what stands out as probably the most remarkable achievement of science is its success. Ours is a scientific age not only because the discoveries of science are among the great accomplishments in the history of the human intellect or because science has transformed our entire environment, but because our ways of thinking and our notions of truth and reality have become largely those of science. The most dramatic outcome of the long interaction between science and other modes of organizing and interpreting experience seems to have been that, while science has transformed the physical setting of our lives and provided an extraordinary amount of new knowledge and exciting new concepts about the physical universe, it has at the same time undermined the authority of and our confidence in the only ways available to us of incorporating these into a significant relationship to our philosophical needs as human beings.

III
Criticism of the Humanities by Scientists

There has probably never been a time when the humanities have not been the object of some kind of criticism. The arts and letters reflect the intellectual, political, and social conditions of their times, and they also comment on them. They do not always conform to what particular individuals and groups would like to hear, and their shortcomings are accordingly made known to them. Sometimes, however, they are criticized for what they are essentially. Plato regarded the poets as a danger to his ideal commonwealth, and at least some theologians during the Middle Ages and the early Renaissance regarded secular literature with suspicion. It is not wholly unrelated to this phenomenon that some of the classics of criticism have been explicitly or by implication defenses of poetry. At the turn of the century, when the energies of most modern nations were preoccupied chiefly with industrial and commercial development, the humanities were sometimes dismissed as impractical. This view still has currency: Howard Mumford Jones in his treatise on the humanities, *One Great Society*, proposes a series of questions which "business leaders would presumably want answered if they were called upon to support scholarship in this field," [1] and many of these questions raise the issue of practical relevance. Today, however, the protest which seems to carry the most weight is that which comes from science. Not only have the development and increased importance of science affected the status of the humanities in our time, but they have also provided a new source of criticism for them and a new body of critics. Now that the scientists have come to play a leading role in our society, they have also become understandably concerned with the present welfare and future progress of their work, which they tend generally to identify with the present welfare and future progress of society; and they are therefore sensitive to any sources of misunderstanding or opposition. Many of them have detected what they believe to be a belligerent or ignorant attitude among artists, literary men, and non-scientific scholars—the group whose collective activities are usually summed up in the word "humanities." Where they thought to have allies, some have found enemies. Others, who have not detected active opposition on the part of the humanities, have found them neglectful of and indifferent to their proper role—a dormant power, uncomprehending and neglectful of the one dominating creative impulse of our culture.

Underlying some of these views is an uneasy awareness of something not quite satisfactory in the present status of science in our culture. Its intellectual achievements are widely acclaimed, its technological developments are the hallmark of our civilization, its vital importance to the preservation of our kind of society is acknowledged; and yet, somehow, there appears to be a disappointing and even ominous lack of harmonious association of this great power and these vast accomplishments with the total activity and outlook of our culture. It is with a statement of this theme that James Bryant Conant opened his Terry Lectures, in 1946:

> Is it not because we have failed to assimilate science into our western culture that so many feel spiritually lost in the modern world? So it seems to me. Once an object has been assimilated, it is no longer alien; once an idea has been absorbed into an integrated complex of ideas, the erstwhile intruder becomes an element of strength.[2]

The question which Conant asks here has become a principal motif in discussions by scientists of the present state of our culture, but it is not often that the question is asked with the same temperateness of tone. In its most characteristic recent form, the question has become a positive assertion concerning a lack of harmony and adaptation between the humanities and the sciences, and the answer takes the form of a charge of failure on the part of the humanities as the discipline whose responsibility it is to assimilate into an integrated body of ideas and attitudes the significant intellectual achievements of our time. Thus has grown up the concern over our divided culture—on the one side, a dominant science which is progressive and looks to the future; on the other, the regressive humanities, which have failed to keep pace with the leading intellectual force of our times and instead keep faith with the dead past:

> Our society is indeed divided between the past and the future, and it will not reach a balanced and unified culture until the specialists in one field learn to share their language with those in another. The scientist has much to learn still, in language and thought, from the humane arts. But the scientist also has a share, a growing share, to contribute to culture, and humanism is doomed if it does not learn the living language and the springing thought of science.[3]

It is, presumably, because they understand the value of the humanities that the scientists condemn their present incapacity. "Professors of science," writes Joseph Gallant, "apologetically admit the straitened confines of technological studies and defer to the deeper wisdom of the humanities. But the humanities sweepingly ignore the role played by scientific insight and thinking in the ideology of our times and disdainfully march on their archaic way as though the atomic and electronic age had not yet arrived." [4]

The most widely discussed statement of this view, and the one which has given currency to the phrase "two cultures," is that of C. P. Snow, *The Two Cultures and the Scientific Revolution*, originally delivered as the Rede Lectures in 1959. What has given Snow's expression of the idea such authority is not only that it has come as a climax to a growing conviction among scientists, but also that he speaks as one who has lived in both worlds and enjoys the unusual position of being both a scientist and elder statesman of scientific activity and also a respected novelist. This is how he sums up the present situation:

> Literary intellectuals at one pole—at the other scientists, and as the most representative, the physical scientists. Between the two a gulf of mutual incomprehension—sometimes (particularly among the young) hostility and dislike, but most of all lack of understanding. They have a curious distorted image of each other. Their attitudes are so different that, even on the level of emotion, they can't find much common ground. . . . The non-scientists have a rooted impression that the scientists are shallowly optimistic, unaware of man's condition. On the other hand the scientists believe that the literary intellectuals are totally lacking in foresight, peculiarly unconcerned with their brother men, in a deep sense anti-intellectual, anxious to restrict both art and thought to the existential moment. And so on. . . . On each side there is some of it which is not entirely baseless. It is all destructive. Much of it rests on misinterpretations which are dangerous.[5]

It is worth looking at the rhetoric of these statements, for it reveals the intensity of feeling, whether in sorrow or anger, which the issue arouses. Gallant: the scientists *"apologetically admit"* their limitations, but "the humanities *sweepingly ignore* the role played by *scientific insight* and *thinking* in the ideology of our times and *disdainfully* march on their *archaic* way." Even in Bronowski's more balanced statement, though the scientist has "much to learn still," "humanism is *doomed* if it does not learn the *living* language and *springing* thought of science." Snow's argument appears on the whole to be impartial, but where the rhetoric is loaded it is against the humanists: "at one pole the literary intellectuals, who incidentally while no one was looking took to referring to themselves as 'intellectuals' as though there were no others." [6] And there is his comparison of Eliot and Rutherford as type figures:

> Non-scientists tend to think of scientists as brash and boastful. They hear Mr. T. S. Eliot, who for these illustrations we can take as an archetypal figure, saying about his attempts to revive verse-drama, that we can hope for very little, but that he would feel content if he and his co-

workers could prepare the ground for a new Kyd or a new Greene. That is the tone, restricted and constrained, with which the literary intellectuals are at home; it is the subdued voice of the culture. Then we hear a much louder voice, that of another archetypal figure, Rutherford trumpeting: "This is the heroic age of science! This is the Elizabethan age!" Many of us heard that, and a good many other statements beside which that was mild; and we weren't left in any doubt whom Rutherford was casting for the role of Shakespeare. What is hard for the literary intellectuals to understand imaginatively or intellectually is that he was absolutely right.

And compare "this is the way the world ends, not with a bang but a whimper"—incidentally, one of the least likely scientific prophecies ever made—compare that with Rutherford's famous repartee, "Lucky fellow Rutherford, always on the crest of the wave." "Well, I made the wave, didn't I?" [7]

When scientists discuss the divided nature of our culture, the humanities are likely to come off badly.

To get at the heart of the problems of the disunity of our culture, it is necessary to go beyond the alleged failures of the humanities and the limitations of the temperament of the present day humanist in comparison with his scientific counterpart, and to take into account certain distinctive features of the learning of our civilization. To a degree far beyond that of any previous age, our intellectual activities in all areas of learning are vast, they are highly specialized, and their results at the most advanced level are unusually recondite. The magnitude of our researches is appalling, not only in the number of persons who continuously devote their time to it and in the quantity of learning which it makes available, but also in the accelerating rate of its accumulation. Moreover, the preservation of the record of the activities of our culture comes close to being total. Every scrap of official, and much unofficial, data is preserved; every successful research effort is published, along with some that have not been successful. To cope with any part of contemporary knowledge in a professional way, one has to select a limited specialized area; to keep up with even a limited portion of it has become an unremitting and exhausting task. Moreover, the state of learning has become so advanced that the extreme boundaries are accessible only to fully equipped explorers. This is as true outside of the natural sciences as it is in them. To the non-economist, for instance, the theory of money, once we get past a few elementary principles, soon reaches levels where the atmosphere is too rarefied for the unconditioned. The exact understanding of such writings as the poems of T. S. Eliot, or Rilke, or the late experiments of Joyce, is almost as far from the capacity of the interested amateur reader as is the proper understanding of Einstein.[8]

A sobering fact about our culture is that, while its total knowledge is vast and of extraordinary refinement, and our capacity to uncover more—and more recondite—knowledge is apparently inexhaustible, the area of individual ignorance is becoming correspondingly greater. This axiom applies not only to the educated man outside a particular field but also to one in it. Every notable advance in some area of physics, for instance, renders not only the non-physicist more ignorant, but also the average physicist as well. In every science we will find good men who modestly concede that, concerning some significant phase of their subject, they do not know enough to have an expert opinion. It is true, of course, that all learning involves exclusion and that anything learned is at the expense of something not learned. But never before has this been so true for the serious, educated man who is eager for access to the new and wonderful things which the advanced learning of his culture is able to provide. For one thing, the sheer bulk is too great, and, what is more to the point, when he reaches for the most exciting of our intellectual products he is likely to discover that they are really beyond his grasp. This is the theme of some reflections by Robert Oppenheimer in the ACLS annual lecture in 1959:

> Common sense and specialized knowledge are in a very special, unsymmetric relation to each other. All our knowledge, all our specialized knowledge, starts with common life: words which we know and do not have to argue about, that are in our experience. Then we begin to manipulate, intellectually and physically; and new things grow. . . . What flows back from special knowledge, back into common knowledge, is rather a small part. I am not talking so

much about the fact that we use difficult words. I am talking about the fact that behind the difficult words there is a difference of experience, in life and tradition, which is very hard to bridge. Anyone who tries to tell you what is going on in the specialized parts of knowledge—and this I believe is not quite as true of the anthropologist as of the physicist, not quite as true of the philologist as of the biologist—has some of the same problems as a man who has been off to war for five years talking to people who stayed at home, or a man who has been in prison; but in addition there is of course the intellectual problem, again varying in difficulty from subject to subject, particularly difficult where the abstractions of an explosively growing mathematics are involved. . . . The deep things in physics, and probably in mathematics, are not things you can tell about unless you are talking to someone who has lived a long time acquiring the tradition.[9]

The general situation in the world of learning is the same for the non-sciences as for the sciences, but, as Oppenheimer's remarks point out, the sciences represent a special case in the whole situation. One advantage of the non-sciences is that our education requires all of us to master the symbolic tool of language and to familiarize ourselves with its operation in a great variety of contexts and forms. Serious writing raises no real barrier if it avoids words which have meanings only for the initiates in a specialty and employs concepts that derive from common experience or the elements of our common education. The most original learning, and also art, of our times is unlikely to fall into this category, and a sense of alienation arises accordingly; but in the nature of things, the most acute sense of alienation will arise from those aspects of learning which demand highly specialized symbolic tools, such as advanced mathematics—in short, those aspects of science which we are told are the glory of our age and have transformed the physical universe in which we live. The separation between the non-scientist and the scientist will therefore be the most striking of all. But to see only this is to miss the essential difficulty. A gulf separates all advanced learning from the curious educated man; and even within the learned world itself, even within the sciences, the character of modern learning cuts men off from easy access to what the best minds in any group have to offer:

> One is faced here [Oppenheimer continues] with a situation in which the practitioners of the specialized sciences have between them contacts, valuable, important; but there is no total relevance, no total mapping of one on the other; and between all these people—and as of now it is still a very small part of our society—between these and the people who do not live in this world, there is only such communication as is mediated by earlier or later education, by friendship, by patience, and by the best of good will. That is why the core of our cognitive life has this sense of emptiness. It is because we learn of learning as we learn of something remote, not concerning us, going on on a distant frontier; and things that are left to our common life are untouched, unstrengthened and unilluminated by this enormous wonder about the world which is everywhere about us, which could flood us with light, yet which is only faintly, and I think rather sentimentally, perceived.[10]

If the present alienation between important areas of learning is ever to be removed and new learning made a part of the common intellectual experience of all educated men, we will have to find some way out of the accelerating accumulation of new knowledge, and the specialization, refinement, and strangeness of the advanced research which is the glory and burden of our civilization. In this situation, the position of the humanist with reference to science is the most difficult of all, and if his involvement in science is less than enough, it should not, at the very least, be attributed to indifference to the claims of science and ignorance of its importance. There is evidence of alienation on the other side without the same basis for lack of comprehension. Snow observes that, in the course of interviewing thousands of well-trained and intelligent scientists and engineers, he discovered that only a few of them had any first-hand experience with works of literature, familiarity with which would be taken for granted by literary people. Much of this literature is not beyond the reach of an educated man. Perhaps there is something in the very nature of our specialized learning

which creates alienation even where the ability to comprehend is not involved. We can understand that a young
scientist, confronted with the great body of knowledge and
the numerous devices and techniques which he is called
upon to master, subject to great pressures to make discoveries
and excel, and convinced that at this moment in history
what he is doing is the most important thing of all, will feel
reluctant to devote any time to the novels of Dickens or
Faulkner and the poetry of Chaucer or Eliot. But even more
important than this is that it is possible to define his role as
a scientific man without any reference to literature, except as
a personal ornament or relaxation from his intense duties.
It is not, however, possible to define the role of the humanist
in such a way as to exclude preoccupation with what the
scientists are doing, for the humanities lose their power to
the extent that they do not interpret the genius of their age;
and our age, though it is many things in addition to being
scientific, cannot be adequately characterized without taking
science and the scientific spirit into account.

For this reason, what the scientists might have to say about
the arts has some bearing on the whole problem of our divided culture and the responsibility of the humanities. On
this issue Snow is blunt:

> It is bizarre how very little of twentieth century science
> has been assimilated into twentieth century art. Now and
> then one used to find poets conscientiously using scientific
> expressions and getting them wrong—there was a time
> when "refraction" kept cropping up in verse in a mystify-
> ing fashion, and when "polarized light" was used as though
> writers were under the illusion that it was a specially ad-
> mirable kind of light. Of course, that isn't the way that
> science could be any good to art. It has got to be assimilated
> along with, and as a part of, the whole of our mental ex-
> perience, and used as naturally as the rest.[11]

The response of the arts and literature to science is represented as ill-informed and trivial. Bronowski appears to
contradict this view. He contends that "a civilization cannot
hold its activites apart" and that it "is bound up with one
way of experiencing life," and he illustrates the point from
painting: "The study of perspective in the Renaissance

chimes with the rise of sensuous painting. And the distaste of painters for naturalism for fifty years now is surely related to the new structure which scientists have struggled to find in nature in the same time." [12] The example, however, is from the visual arts. Elsewhere, as we have seen, Bronowski speaks with as much feeling as Snow about the failure of humanities. When the scientific apologists criticize the arts of our day for their failure to interpret science or make better use of its new insights, it is literature they have in mind primarily. Thus, Joseph Gallant contrasts the effective way in which Pope, Shelley, Tennyson, Whitman, Emerson, Sandburg, and Hart Crane absorbed and interpreted imaginatively the science of their day, with the almost complete absence of interest among current poets, except for a few "glib satirical pieces [which] have appeared in magazines, chiding science for man's interference with God's ways." [13] Where the modern writers have responded, the results have been trivial (we are not amused).

There is first of all a question of fact to consider—whether, that is, the situation is as these writers describe it. The embarrassing ineptness of those writers to whom Snow refers, when "refraction" kept cropping up in verse in a mystifying fashion, the few glib satirical pieces—those admittedly do look pitiful, if that indeed is the record. It is true that most modern science is not open to the untrained individual, but one might well have expected more. When we look at certain obvious possibilities, the record is only slightly better. The theme of man's position against the staggering picture of the scientific universe, so frequently urged in popular works on science, has not found a response among the poets, although here they might well point out that the seventeenth and eighteenth centuries responded to this theme so thoroughly and expressed it in so many different ways as to leave little in the way of novelty for the modern writer. Novelists have not, to any appreciable degree, found their materials in the world of scientific activity nor modelled their characters on the scientist of our times; though here again it is possible to point out that only recently has the scientist become a public figure and placed himself in the field of public action in such a way as to create interesting possibilities for fiction. The conditions and possibilities of a technological society have not been dealt with extensively in fiction since H. G. Wells explored the field and then expressed his disillusionment with his earlier optimism. And the vast amount of science fiction has not so far risen above the level of the best detective yarns. Yet all these might be considered matters of surface importance. We must agree with Snow that, if science is to be any good to art, "it has got to be assimilated along with, and as a part of, the whole of our mental experience, and used as naturally as the rest." It may therefore be that we are not at the moment in a position to assess accurately the manner in which the science of our century has left its impress on the important writings of our times. What we know about the influence of science on writers of the past has been made apparent chiefly as a result of a considerable amount of scholarship during the last few decades—a fact in itself somewhat at variance with the notion that humanistic scholars are not interested in the effects of science on our culture. Nothing on the same order of thoroughness has been done, or could in fact have been done, for the scientific influences on the literature of our own day, and accordingly we are not in so good a position to generalize about the matter as we are about earlier European and American literature. Nevertheless, we know enough about the art and literature of our day to make a few suggestions.

If we consider the spirit of modern science rather than its theories and their implications, we will discover a surprising analogy between the development of art and literature and that of science within the last century. Those who look for an immediate and direct response to the discoveries of science find only triviality, timidity, and staleness in the work of modern writers. Snow's picture of T. S. Eliot as the archetypal figure of modern literature, timidly suggesting some modest hopes for a revival of verse drama, can be paralleled by statements like that of Gallant, who says of modern literature that "the revelations and moods presented are trivial variations on themes centuries old." [14] Whether *Ulysses* is a trivial variation on *Tom Jones,* or *Four Quartets* on Donne's religious poetry, may with propriety be regarded an open question. The art and literature of the last century reveal, even in the most superficial review, continuous experimentation, some of it highly original and productive of permanently valuable results, the rejection of many traditional forms and methods, and the pushing of the boundaries of art in unfamiliar and surprising directions. They have, that is, shown something of the dynamism and revolutionary character of science. They have explored by their characteristic means the universe which is their province, and have cultivated new forms of sensibility and introduced a new order of perception of the physical world. As science has refined its symbolic tools in order to probe further into the physical universe, so artists have experimented with new media and with novel uses of the old, musicians have worked with new tonalities and instrumentation, and writers have experimented with language, forcing it into involved types of symbolism and unusual patterns of construction. They have also devised new or radically modified patterns of formal order. The form of *Tom Jones* is governed by a scheme of time sequence and causality which would not serve at all for Joyce's *Ulysses*, with its interest in the simultaneity of events and its concern not so much with the event as with the unspoken response of the individual's consciousness to it. In a similar way, the Newtonian scheme of order would be unsuitable to a physicist concerned with an order of events that requires Einstein. Whether these new developments in the arts owe anything to analogous developments in science, or whether they are simply independent creations of artistic insight, like the "Freudian" elements in Shakespeare and Dostoevsky, it is premature to say. They must be acknowledged, however, to be more than trivial, and to parallel in their own way the dynamism and revolutionary direction of modern science. And in the long run they may prove useful to science, for they may be a better means of adapting the sensibilities of modern man to an appreciation of the unusual schemes of order which science has been creating than if writers of fiction were to create a vogue for scientists as heroes of novels or if poets were to write rhapsodies on the galaxies, with a more scrupulous attention to a correct use of scientific terms.

It is not likely, however, that these manifestations of

originality and novelty in the arts of our age will be reassuring to the scientific critic of the humanities. He would still be troubled by a characteristic weakness of the humanities which he regards as endemic and which in his opinion sets them apart from the creative spirit of modern science. The humanities, he would say, have an attachment to absolutes, which imposes upon them a chronic inflexibility of approach and so separates them from the scientific attitude and the dynamism of the scientific quest. As Eugene Rabinowitch puts it:

> The scientific approach to the world around us is exploratory, tentative, relativistic, quantitative; it constructs temporary working hypotheses and discards them; it expresses its conclusions in terms of relative probabilities, not of absolute certainties. The humanities, on the other hand, are accustomed—in particular in the study of law, government, and social and political sciences—to qualitative arguments and absolute concepts.

He quotes George Boas in support of this view: "Humanities have remained, on the whole, slave to what their practitioners believe are universal and eternal values." [15] Such views represent the situation imperfectly, and for science as well as for the humanities. It is sometimes forgotten that science, too, assumes, if only pragmatically, certain absolutes and "universal and eternal values." One of these is the assumption of some kind of regularity and order in nature which may be discovered and represented in intellectual terms. The picture of this order is subject to revision, but on the belief in the existence of some regularity and order depends the very idea of science. Scientists also insist on rigorous "proof" and the elimination of "error," in short, on truth. They have a preference for "elegant" rather than overcomplicated solutions, other things being equal. The acceptance of these absolutes and values does not prevent the sciences from being "exploratory, tentative, relativistic." There is also some misunderstanding involved in the assumption that the humanities are rigid in their adherence to absolutes and thus lack the realism and flexibility and adaptability to new facts which are the virtues of science.

The humanities are concerned with man and the human condition. They are therefore preoccupied with the choices which are open to men, the goals men set for themselves, and the means they choose to gain them; and they are consequently further preoccupied with how these affect the character and the quality of an individual's life, and to what extent and in what ways they make possible the full realization of the capacities which appear to distinguish men from other forms of sentient life. The humanist is therefore deeply concerned about qualitative matters. Rabinowitch contrasts the quantitative character of science, to which he gives favorable emphasis, with the "qualitative arguments" of the humanities, an attribute which carries pejorative implications. Scientists, as men, are also interested in qualitative matters. They ask for freedom of inquiry, they have esthetic interests, and many of them have become commendably concerned about the implications of their discoveries for the welfare and happiness of mankind. Such matters are not, however, the province of their study as scientists as they are the area of inquiry for humanists. Quantitative results are more readily obtained and can be more exactly formulated than qualitative generalizations, but the observations supporting the latter do not need to be less realistically conducted or less honestly set down, or thought about with less discipline of mind than the scientific. They may be, and often are, but there is no justification for insisting they must be so in principle. The humanist's field of observation and study being what it is, he cannot escape becoming involved with values, with such considerations as justice, goodness, beauty, happiness, and the like. He is a "slave to absolutes," however, only in the sense that the scientist is a slave to the absolute of regularity and order in the universe which he continually seeks to discover. There is no one way in which these values may be defined or represented as being expressed and realized in action. Humanists have never adhered to any single unvarying "hypothesis" about their nature, and they have invariably subjected these absolutes to new scrutiny as changes in the human scene placed them in a new light. To this fact, the history of philosophy, literature, and the arts testifies. The law, which Rabinowitch cites in particular, affords many illustrations of conscious efforts to realize the absolute of justice with due consideration for contingency and the continual changes of the human scene. We need only recall the interpretation of the "public welfare" and "due process" clauses of the Constitution of the United States by successive decisions of the Supreme Court as an example of the efforts of serious-minded jurists to preserve the vital meaning of a basic legal document, intended to preserve justice in a well-ordered state, by accommodating it to changing social, economic, and technological conditions. The humanities are not contrary and opposed to the spirit of science by virtue of their qualitative approach and their concern for values, but they are different from science in these respects and their role is complementary to that of science.

What the humanities have produced in the exercise of their role in our day has not found a universal welcome among scientists. It has been observed that their preoccupation with the human condition and the individual experience has led them usually to a tragic sense of life and in consequence to indifference toward the possibilities for the amelioration of man's lot in general. Snow contrasts the scientist's useful optimism and the humanist's defeatist tragic view:

> Most of the scientists I have known well have felt—just as deeply as the non-scientists I have known well—that the individual human condition of each of us is tragic. . . . But nearly all of them [scientists]—and this is where the colour of hope genuinely comes in—would see no reason why, just because the individual condition is tragic, so must the social condition be. Each of us is solitary: each of us dies alone: all right, that's a fate against which we can't struggle—but there is plenty in our condition which is not fate, and against which we are less than human unless we do struggle. . . . There is a moral trap which comes through the insight into man's loneliness: it tempts one to sit back, complacent in one's unique tragedy, and let others go without a meal. As a group, the scientists fall into that trap less than others. They are inclined to be impatient to see if something can be done, until it is proved

otherwise. That is their real optimism, and it's an optimism
that the rest of us need badly.[16]

This statement desperately calls for some clarification, for the
distinction is neither so simple nor so self-evident as Snow
makes it out to be, and there is reason to believe that it is
seriously confused if not simply wrong.

There are two aspects to science: one is science as an intellectual adventure of the highest sort whose sole justification
is the exercise of man's mind upon the world of physical
phenomena; the other is science as a power which has extended man's control over nature and has produced a trans-
formation of man's physical environment. In the present
effort to describe the scientist's place in our culture and to
promote his activities, both have been emphasized, depending upon which one is required by the argument. The two
views of science have fluctuated in importance during the
history of science. Bacon considered the aim of science to be
the improvement of man's lot, and he accordingly made
charity the *primum mobile* of scientific endeavor, and his
ideal scientist was motivated by a sense of pity for man's
condition. During the intervening years, the scientific ideal
moved away from Bacon's conception and the disinterested
search for truth replaced it as the motive and justification of
the scientific adventure at its best. The motives of individual
scientists vary, and there is also some variation between expressed motives and those which remain unspoken, perhaps
because the individual is not even fully aware of the latter.
In general, however, the more advanced, theoretical, and
original the science, the less likely is it to be justified today
on the grounds of utility, except as an argument for financial
support. In striking contrast to his sentiments, quoted above,
about the scientist's social conscience and social optimism,
Snow himself recounts how he and his fellow students of
science at Cambridge "prided ourselves that the science we
were doing could not, in any conceivable circumstances,
have any practical use. The more firmly one could make that
claim, the more superior one felt." [17] Snow is not sympathetic
with the scientists for their usual lack of respectful interest
in the work of engineers, but the attitude he describes can

be duplicated in any laboratory engaged on what is referred to as basic research. The story of the beginnings of the Royal Society could, in fact, be thought of as a fitting archetypal myth emblematic of the character of the scientist and his endeavor. The small group of men who attended the informal meetings at Gresham College did so because they found, in this congenial activity and association, a welcome escape from the prevailing religious controversies and political struggles through their participation in fascinating problems which were unrelated to the deepest issues of their times. The world in which the scientist is, by virtue of his education, trained to operate in most effectively is an ideal world submissive to intellectual control and beyond good and evil. Excellent bacteriological research has been done irrespective of whether support for it came in the interest of the wine industry or of bacteriological warfare. Snow warns of the danger of complacency in the humanist's sense of one's tragic loneliness, but there is at least equal danger of complacency in the scientist's primary concern with problems which are dissociated from immediate human concerns or their possible effect on the lives of men and which find their justification as adventures in discovery or as pure intellection. Snow notes that the culture of the young scientists he has interviewed "doesn't contain much art, with the exception, and important exception, of music." [18] This susceptibility to music is not surprising. Music is the most ideal of the arts, the most abstract. It is at the other pole from literature, in which preoccupation with the human condition in all its manifestation and life values in all their variety is essential. As individual men, scientists will manifest different degrees of concern for the social condition, but, Snow's arguments to the contrary, scientists as a group have not shown any singular propensity for worry over whether "others go without a meal."

The optimism of science has no necessary connection with compassion and a desire to improve the lot of man even in the case of the Baconian ideal. It is both a condition of and a consequence of the scientist's activity. It arises from the capacity of science to isolate a problem, set it within an investigative framework which allows for the exclusion of any primary concern for its human involvement, and then apply various appropriate methods until an acceptable solution of the problem is arrived at or its possibilities exhausted. Experience has shown that, although at the highest level success is elusive, as it is in all forms of human endeavor, at the level of normal operation the probabilities for success in scientific work are very high and the supply of problems that will lend themselves to these methods and provide gratifying solutions is practically endless. The optimism of the scientist is different in origin, and hence different in kind, from that which is demanded in the contingent world of human affairs, and it is questionable, therefore, whether it can be transferred to those who bear the burden of decision there. An error of judgment on the part of a scientist will usually require no more than a reformulation of the problem, and a mistake on his part will normally require the repetition or redesigning of the experiment; in the case of a statesman or even an industrialist, an error may result in either immediate disaster or serious human dislocations. Optimism in such endeavors is of a different brand from that which is inherent in the scientific activity. It is also different in the case of a nurse working in a wretchedly poor and hopelessly overcrowded tenement district, or the chairman of a city crime commission continuing year after year in the face of a corrupt police system. And just as there is no necessary correlation between the optimism inherent in science and a concern for the amelioration of man's lot, which science has been instrumental in promoting, so there is no necessary correlation between a tragic view of the individual life and complacent indifference toward possible improvement of the social condition. It is, in fact, difficult to come up with names of scientists whose activities have been as strongly motivated by a concern for the social condition as, for example, Dickens— certainly not Snow's archetypal figure, Rutherford.

That science has contributed lavishly to the amelioration of the condition of men, beyond the Baconian dream, is one of the things which scientists have a right to mention with pride, but the passion and drive to improve man's lot has come from political leaders, social reformers, missionaries, industrialists on the lookout for new useful products, some "applied" scientists and medical men, and even cranks, and not the scientists Snow is writing about. They would, in fact, have served mankind less well had they been motivated otherwise than they were and are. Conant refers to science as "strengthening the hand of the Good Samaritan." "This consequence of science," he continues, "needs to be underlined. If loving your neighbor as yourself is the epitome of a religious outlook, it can only have meaning as a policy to the extent that one is able to help the neighbor when he or she is in pain or trouble." [19] The good Samaritans of our civilization owe gratitude to science for strengthening their hand, though they themselves are likely not to be scientists. And of course there are times when the help the neighbor needs is not bread, and when his pain does not lend itself to the ministrations of science.

What appears as hostility or indifference to science or a failure to be sufficiently aware of its importance is sometimes quite the reverse of these, and represents a preoccupation with the implications of the dominant position of science today and of certain ways in which science has changed our world. The almost universal acceptance of science and scientific procedure and of the scientist's approach to reality has placed the humanistic approach to experience in a dubious position without at the same time rendering it irrelevant or replacing it with a useful substitute. Science has in some respects had a liberating effect on men's minds and lives, but the conditions created by our scientific and technological civilization have also facilitated the increase in control by centralized authority and thus aided in depressing the freedom and independence of the individual. Moreover, in the half century during which science and technology have made their greatest strides, the world has been threatened with disasters—devastating wars, serious economic calamities, revolutions and violent social changes, and systematic, efficient inhumanity on a scale that travesties our claims to progress and civilization. Any elation over one development is more than qualified by dismay over the other, and though science cannot be identified as the cause of these disorders, its fruits have undeniably increased the power and destructiveness of the elements which create them. Science as a daring enterprise of the mind and spirit shares in the universal respect which is accorded to all creative minds, all men of learning, who are partners in the great human adventure of knowing. Science, however, is also a power, which contains no inherent imperatives for its proper human use, and in this capacity modern science has presented mankind with problems which might well be beyond our capacity to solve.

These various considerations complicate the view of science for the humanist. He cannot separate the whole phenomenon of modern science from the issues which are basic to his approach to experience and he cannot, therefore, preserve the simple uncluttered view of science which arouses the unqualified esthetic and intellectual enthusiasm of the dedicated scientist. And when he reflects on the disunity of our culture, it is likely to have a meaning for him which would not readily occur to scientists.[20] His attitude arises, moreover, not from an indifference to modern science or a lack of appreciation of its brilliance and importance but, on the contrary, from a realization of the nature of science and its place in the modern world. The serious writer and artist are bound by the same demands for honesty and integrity in their search for truth and reality as the scientist. The tragic overtones sometimes discernible in their work are a consequence of their candid response to the world as they find it, including the world of science, and it is surely through some profound lack of understanding that the appearance of the tragic sense in the modern writer comes to be labelled "complacent."

It is possible to admire science, to enjoy the comforts afforded by modern technology, and to be grateful for the freedom from want, pain, and slavery which these can bring, without accepting a confident and enthusiastic view of present prospects and rejoicing over what man has wrought. If poets today do not, like Addison, write odes with a scientific coloring to the spacious firmament on high, or rhapsodize, as did Thompson, over the heroes of science and the universe they had revealed, it is because some of the foundations for the early confidence and elation have been undermined in the meantime. Snow compares the timid voice of "the culture" with Rutherford trumpeting, "Well, I made the wave, didn't I?" One may admire Rutherford without sharing in his exultation, for the wave is beginning to look ominously tidal. The gifts of science are lavish, and all science asks for is a little appreciation, plenty of support, and the opportunity to provide even more lavishly, but the giver is beginning to act compulsively and some of the benefactors are beginning to wonder if they are not about to be killed with kindness. And so the non-scientist, though not necessarily a primitivist, is inclined to question the scientist's simple faith in the supreme value of untrammeled scientific activity and its accompanying accelerated technological progress. W. H. Auden, for example, writes:

> We have all accepted the notion that the right to know is absolute and unlimited. The gossip column is one side of the medal; the cobalt bomb is the other. We are quite prepared to admit that, while food and sex are good in themselves, an uncontrolled pursuit of either is not, but it is difficult for us to believe that intellectual curiosity is a desire like any other, and to realize that correct knowledge and truth are not identical. To apply a categorical imperative to knowing, so that instead of asking, "What can I know?" we ask "What at this moment, am I meant to know?"—to entertain the possibility that the only knowledge which can be true for us is the knowledge we can live up to—that seems to all of us crazy and almost immoral.[21]

For those who think this is stuffy and as irrelevant as the forbidden tree in the Book of Genesis, we can provide a somewhat different version of the same notion, this time by a biologist:

> Now it can be said that it is possible to achieve almost anything we want—so great is the effectiveness of technology based on the experimental method. Thus, the main issue for scientists and for society as a whole is now to decide *what* to do among all the things that could be done and should be done. Unless scientists are willing to give hard thought—indeed, their hearts—to this latter aspect of their social responsibilities, they may find themselves some day in the position of the Sorcerer's Apprentice, unable to control the forces they have unleashed. Any they may have

to confess, like Captain Ahab in *Moby Dick*, that all their methods are sane, their goal mad.[22]

And for those to whom even this seems vaguely moralistic, we can recommend the analysis by Derek Price of the growth of scientific knowledge and scientific activity and its implications for the state of affairs in the near future. It is the mathematical counterpart of the allusion in the previous passage to the Sorcerer's Apprentice. It may be possible for scientists to dismiss Auden; it is another matter when an enthusiast for science writes, as Price feels compelled to do in the light of his graphs and figures, "We must not expect such growth to continue, and we must not waste time and energy in seeking too many palliatives for an incurable process. In particular, it cannot be worth while sacrificing all else that humanity holds dear in order to allow science to grow unchecked for only one or two more doubling periods." [23]

To focus, as Snow seems to do, on a disabling preoccupation with the individual death as the source of the modern humanist's tragic view, in contrast with the scientist's future-directed and meliorist attitude, is certainly to oversimplify, if not to misunderstand. It is, after all, not quite so important that each of us dies alone as that each of us must live with himself and his fellows. Rightly or wrongly, many serious writers of our times have expressed grave doubts that the world which we have created is compatible with the needs of the spirit of man. From their particular angle of observation, they are unable to entertain the same view of the prospect before us as the average man of science. Commenting on the development of the modern attitude toward scientific knowledge during the last three to four centuries, Erich Heller writes:

> In the course of those centuries the poetic truth of [Marlowe's] *Dr. Faustus* has been rendered into the prose of science; and in the process it has shed all theological inhibitions fostered by the morality of the old Faustian plot —the morality of the Tree of Knowledge. The serpent has been chased off its branches, and the tree, bearing sinful fruit no more, received, on the contrary, its glorification at the hands of the new age. The searching mind and the restless imagination were declared sacrosanct. It was a stu-

pendous revolution, glorious and absurd. Its glories need no recalling. They still lie in state in our universities, our theatres, and our museums of art and science.

But its absurd consequences pursue us, alas, with keener vivacity. For we make a living, and we shall make a dying, on the once triumphant Faustian spirit, now at the stage of its degeneracy. Piccolo Faustus has taken over the world of the mind. Wherever he sees an avenue, he will explore it—regardless of the triviality or the disaster to which it leads; wherever he sees the chance of a new departure, he will take it—regardless of the desolation left behind, He is so unsure of what *ought* to be known that he has come to embrace a preposterous superstition: everything that *can* be known is *worth* knowing—including the manifestly worthless. Already we are unable to see the wood for the trees of knowledge; or the jungle either. Galley-slaves of the free mind's aimless voyaging, we mistake our unrestrainable curiosity, the alarming symptom of spiritual tedium, for scientific passion.[24]

The scientific critics of the humanities cannot be expected to be comfortable with these differences in point of view, and it would not be surprising if some present day scientific Plato decided that the poets have no place in the scientific commonwealth. At the end of an essay in which he warns the humanities and social sciences that they "must be permeated with the knowledge and spirit of science if they are to be more than relics of a buried age," the biologist, Bentley Glass, concludes:

> We cannot endure, half scientific, half rebelliously non-scientific. The schizophrenia of the "two cultures" leads only to disaster. As Bertrand Russell has so well said, science can enhance among men two great evils, tyranny and war. And which would be preferable, do you think, to perish in a nuclear holocaust or to live under a scientific tyranny.[25]

The effect of this remarkable statement is what comedians refer to as a double take. At first one is shocked at the choice which is presumably left us. On second thought, it becomes more shocking that the writer has expressed the choice in the form of a rhetorical question, as though it is self-evident that there is only one obvious answer. But on further reflection it becomes equally astonishing that no third possibility is offered, as though the problem of first magnitude were not to exhaust all our human resources to avoid the stated alternatives. Perhaps, however, the chill rigor of the dilemma with which the statement confronts us is nothing more than a desperate rhetorical device to shock us into a realization of our common danger. It is reassuring that in a similar statement in his book, *Science and Liberal Education*, the same writer mitigates the rigor of the choice and adds, hopefully if somewhat ambiguously, that "the arts, humanities, and social studies . . . must mollify, enrich, and protect the sciences." [26] The choice between annihilation or survival under a scientific tyranny is, after all, a barren one. For one thing, a scientific tyranny could not eliminate the danger of perishing in a nuclear holocaust, a danger which exists and will continue no matter what political organization one may choose or be forced to accept. It may be possible to purchase physical survival by simply submitting to any who want what we have and threaten us with force and thus eliminating the motive to conflict. But even if scientific tyranny became universal and world wide, the purchase of survival by the unremitting payment of the tribute of submission would continue, and hence the threat of disaster would not end.

Survival, however, in any reasonable human terms, means something more than continued existence. At the very least, the idea of a civilized society cannot be separated from law, political theory, philosophy, and literature, and to give it any depth we would want to encourage the perceptions of the physical world provided by art, the sensibility cultivated by music, and the intellectual excitement provided by the scientific investigation of nature. But however one would describe the requirements of a decent human society at this stage of human development, it would include something more than the preservation of life for some and the opportunity to conduct scientific research for others. Like any other form of tyranny, a scientific one would have to stifle all other forms of creativity. As Conant observes, "A dictator wishing to mold the thoughts and actions of a literate people could afford to leave the scientists and scholars alone, but he must win over to his side or destroy the philosophers, the writers and the artists." [27] Conant erred in excepting all scholars: a dictator cannot endure honest historians and literary critics. A scientific and technological civilizaton cannot endure without a flourishing scientific community, but scientific knowledge and activity cannot insure the well-being or even the survival of such a civilization. A scientific tyranny could survive only if supported by a political tyranny. It is unlikely that in its pure form scientists would have much taste for it.

A severely critical view of the humanities is not shared by all scientists, nor do all scientists accept the radical conclusions which some of their fellow scientists feel obliged to draw from the notion of a science-centered culture. The report of the President's Science Advisory Committee, *Scientific Progress, the Universities, and the Federal Government* (November 15, 1960), recommends, in fact, a balanced and proportionate development of all forms of creative activity required by a cultivated society:

> Obviously a high civilization must not limit its efforts to science alone. Even in the interests of science itself it is essential to give full value to and support to the other great branches of man's artistic, literary, and scholarly activity. The advancement of science must not be accomplished by the impoverishment of anything else, and the life of the mind in our society has needs which are not limited by the particular concerns which belong to this Committee and this report. . . . Neither the government nor the universities should conduct the support of scientific work in such a way as to weaken the capacity of American education to meet its responsibilities in other areas. The cost of scientific progress must not be paid by diverting resources from other great fields of study which have their own urgent need for growth.[28]

This is an admirable statement, but it is far from evident that scientists as a whole are fully aware of the consequences of urging equal support of all forms of creative and scholarly effort while at the same time calling for extraordinary increases in the scale and tempo of the scientific activity, or why it is that a complex society requires the free and vigorous activity of social scientists and humanists. If they did, they would be more effective as critics of the humanities. Today, the humanities are in need of a searching revaluation of their role, and it would be reasonable to suppose that one of the most useful sources of criticism would be those learned men trained in a field which has become celebrated for its objectivity and impartiality. Anyone looking with such hopes into the writings of scientists that deal with the failings and limitations of the humanities will experience disappointment. There is too much bias, too little knowledge and understanding of the subject, and too little sophistication of argument. The chief merits of this critical literature are that it is sharp, outspoken, and provocative, and these qualities, coupled with the prestige that is associated with anything that comes from that quarter, have brought wide attention to the scientists' criticism of the humanities and rendered it unusually persuasive. The judgments supported by these writings are fast becoming the clichés of the current intellectual chatter. The criticism of the humanities by the scientists thus proves to be doubly frustrating. Because of its intellectual shortcomings it cannot be very useful to those who might otherwise have profited from it, and because of its popular success it burdens the humanities with the distracting and unamiable necessity of rebuttal.

IV
The Scientists and Public Affairs

Science has been steadily coming into its own since the scientific revolution of the sixteenth and seventeenth centuries, but a full awareness of the role of the scientist in the modern world is of fairly recent origin. It is true that his value to a very advanced industrial society was steadily being recognized, but this recognition was gradual, and in comparison with the events of the last three decades its progress was glacial. The scientist owes the rapid enhancement of his eminence largely to his great services during the last war—a role only faintly foreshadowed by the work of scientists during the First World War—and to his importance in the continuing power conflict between two political colossi. It is acknowledged that he is essential to our survival which, paradoxically, his gifts have rendered so perilous.

How intimately science has become involved in the affairs of the modern world is illustrated by the curious turn which has been given to the political meaning of a nation's scientific potential by the satellite programs and space research. It is no longer simply a matter of the immediate long-range military significance of a nation's scientific achievements. Whatever military implications the space and satellite programs may have, they are for the moment significant primarily as conspicuous public demonstrations of the originality, strength, and technical knowledge and experience of the scientific and engineering resources of the competing countries, and are thus a means of inspiring home morale and patriotism, reassuring allies, and impressing uncommitted peoples and the enemy—two giants flexing their muscles and performing feats of strength before an astonished world. Indeed, as we contemplate the society which is emerging today and try to define it in terms of its essential needs, wishes, and extravagances—in short, its essential character—we must conclude that the twin culture heroes of our times are the scientist and engineer, for without taking their activities into account it is impossible to define the kind of world in which we live or distinguish our culture from others. Moreover, they alone possess the technical knowledge upon which many of our decisions must depend. "The most conspicuous aspect of our civilization," writes Joseph Gallant, "is the pervasive and ramifying impact of science in every department of life, from household management to warfare." [1] It is difficult to disagree.

As a natural consequence of their commanding role, scientists are rapidly acquiring the principal characteristics of a dominant social class. They enjoy great prestige, they are increasingly called upon to advise industrialists and statesmen, their professional ideals are widely publicized as admirable forms of the private and social virtues, and they are receiving tremendous resources for carrying on their functions. In 1960, an estimated fourteen billion dollars, most of it federal money, was spent in the United States for scientific research and development, an increase of approximately ten per cent over the previous year; and of the total expenditure in 1959 for higher education, an estimated one-fourth represented the amount contributed by the federal government for scientific research and training.[2] All indications are that such expenditures will increase. The involvement of scientists in national affairs has become an accepted and formalized matter, illustrated in the United States by the establishment of the President's Science Advisory Board. Similar arrangements exist in other countries: in England, the Advisory Council on Scientific Policy; in the Soviet Union, the Soviet Academy of Sciences. In the United States, it has also been proposed that a cabinet post be created for science. Within a matter of hardly more than two decades, the scientists have found themselves in a position quite different from that which was normal for their profession before the political upheavals of our century. From a state of modest expenditure and improvisation, they find themselves handsomely provided for in their research, and even their personal financial rewards have improved and are improving. From being left out of consultation on matters which affect them and the progress of their science, they are now eagerly consulted. And with provisions being made for the rapid expansion of scientific and technical training, scientists will shortly be in the majority in the teaching profession and thus in a position to determine educational policy. No class or profession has ever had greatness thrust upon it in so dramatic a fashion. With greatness, however, have come the responsibilities and the somber reflections which accompany it.

There was scarcely any portion of the war effort in which scientists were not somehow engaged, but the episode which more than any other single event signalized the change from the old world to the new was the project to release the energy of the nucleus and harness it for military uses. The success of this project was a brilliant demonstration of the ability of scientists to translate a highly theoretical and recondite statement about the nature of matter into practical application, a feat which some physicists believed to be unlikely. It demonstrated, too, that many individual scientists and engineers could effectively combine their separate efforts under an ambitious and daring master plan, solve numerous problems of extraordinary difficulty, and bring the whole project to a successful conclusion in a remarkably short time. The idea of the solitary scientist working with dedication and patience in his laboratory, while not rejected, was becoming outmoded by the idea of group research. The enterprise was also secret, and thus a departure from the tradition, established by several generations of great scientists, that good science was an international enterprise of free men working under conditions of freedom of choice and of communication.

In some ways the most important feature of all was that the project brought science and government into inseparable partnership. For one thing, the scale of the project was so vast that no single group of scientists, no single research facility, no single industry—for that matter, no spontaneous combination of any of these—could have acted as the organizing and directing force of such a plan or found the resources to finance it. Moreover, the value to the nation of scientific research adequately supported had been so effectively demonstrated that the association of science and government on a permanent and continuing basis became inevitable. There was something unprecedented also in the nature of the results, for there was now made available power on such a scale that some scientists who realized what it might mean hoped that the enterprise would fail. It was not only that this research had resulted in a powerful weapon. It was also that the use and development of power of such magnitude could never be completely divorced from government supervision or public policy. It meant, too, that the scientist could not remain detached from the spheres of action where decisions were taken which involved knowledge of matters about which only he is expert.

Nothing would ever be quite the same again. With the Manhattan Project, scientists left the age of innocence. For many, the principal issue which had been raised was that of the moral responsibility of the scientist. Science was morally neutral, but could the scientist be? The scientist could now with less assurance maintain that the main and sole justification of his work lay in its spiritual, intellectual, and esthetic rewards. The purest research had proved to be the most appallingly practical, and the most tainted. There were those who maintained that the scientist as scientist owed allegiance to only one ethical imperative, to search out the truth about the physical world, and that it was his first moral obligation to carry his researches as far as they would go no matter how pregnant with disaster the results. At the other extreme, some felt that all those, including Einstein, who saw the possibilities for destructive use in the theory of the transferability of mass and energy and who urged its practical application and gave their knowledge and talents to that end were morally wrong and morally culpable. There were also those who believed that scientists cannot avoid their moral obligation to seek the truth as scientists, but that they also have a unique moral obligation to reflect on the implications of what they have discovered, to inform others, and to search for answers to the problems raised by the new powers of science. For some scientists the main issue was the threat to the traditional freedom of scientific activity, not only through the requirements of secrecy but through the direction of scientific research by the needs of government. They viewed these developments as a restriction upon the independence of the individual scientist, and in consequence as a threat to the welfare of science. Scientists found that they were faced with many issues not scientific in nature, though scientific in origin, for which traditional scientific attitudes were inadequate preparation.

These questions affected the individual scientist as a private man, but in view of the public importance of what he was doing they also had implications for his role in modern society. Even the most fastidious moralist and individualist among scientists today realizes that the interest of government in science affects what he will work on and how it is to be supported. On the other side, the government needs the advice of scientists in making decisions that affect its military and diplomatic activities, and the scientists will want to participate in these not only because of concern for their country's welfare but also because, with government as the principal source of financial support for science, what government does will have a bearing on what kind of science will be undertaken. The basic moral and political issues now confronting the scientist have been there for some time, but in former days they could be comfortably disregarded. The bomb dramatized them and made it inescapable that they should be confronted, and gave them an urgency that made them seem new.

They were, of course, not new. They had, in fact, occurred to Bacon. Since he thought of the improvement of man's condition as the first motive to scientific study and since he was enthusiastic and optimistic about the success of science in winning back the lost Paradise, he gave little prominence to the disturbing implications of his plan. Nevertheless, they crept in. Men, he realized, could make foolish and wicked use of the gifts of science, but was this sufficient cause for withholding them? There were dangers in these new powers made available to men, but Bacon expressed hope that religion and right reason would insure their proper use. But in the scientific utopia which he described in *New Atlantis,* he made sterner provision against the abuse of scientific knowledge, as explained by the director of the central research institute: "We have consultations, which of the inventions and experiences which we have discovered shall be published, and which not: and take all an oath of secrecy, for the concealing of those which we think fit to keep secret though some of those we do reveal to the state and some not." In Bacon's ideal commonwealth, secret science is an accepted condition, not merely in the interest of military advantage and national security, but of the general welfare of society. Moreover, in decisions affecting the use of scientific discoveries, the scientists recognize no sovereignty above themselves. Significantly, Bacon realized that the problem would ultimately become one of public policy. However, none of his solutions would have much appeal today. The state as a scientific oligarchy is a remote possibility, and Bacon himself seemed uncertain about the influence of religion and right reason alone as a means of control. The issues, however, remain much the same. Science and public affairs are inextricably related. Decisions will have to be made—many of them. Who will make the decisions? How can we make sure that the decisions will be well informed, sensible, correct? How much of the national income should go to scientific research and development? What areas will receive support? Which will be given priority? The list could easily be extended, along with a supplementary list of the questions and problems raised by any given answer.

The present relationship of the scientist to society involves many serious and pressing matters, most of which are beyond the scope and limited concern of this study.[3] There are, however, some issues arising out of the current discussions of the relationship of the scientist and the non-scientific political administrator that do have a bearing, if only indirectly, on the distinctive character of science and the humanities.

One aspect of the present situation that distresses many

scientists is that decisions on scientific matters which involve the safety of the nation as well as the distribution and use of its resources are left to men who are not scientists. Today there exists an elaborate and at least workable practical system of advising which insures that the opinion of qualified scientists will be consulted on all matters which have to do with science and development, whether it is a matter of weapons, the training of scientists, public health, or the economic implications of new discoveries. It is doubtful whether any scientist believes that the present arrangements are entirely satisfactory. Administrative difficulties aside, they feel that where a knowledge of science is lacking, as it is among most government officials, there can be no proper appreciation of the nature of the problem and hence no basis for a responsible decision. Commenting on the dangers of secret science, where the difficulty can be acute, C. P. Snow writes, "the special aura of difficulty and mystery about these choices will at least be minimised as soon as all politicians and administrators are scientifically educated, or at any rate not scientifically illiterate." [4] But he dismisses this possibility as an ideal solution that is at present quite out of sight.

It seems self-evident that, as scientific matters enter with increasing importance in the affairs of the nation, political leaders must be at a disadvantage if they do not possess some familiarity with science and if they do not have an appreciation of the possible implications of scientific discoveres and technological developments. There can surely be no serious disagreement with Conant's statement of the issue:

> Because of the fact that the applications of science play so important a part in our daily lives, matters of public policy are profoundly influenced by highly technical scientific considerations. Some understanding of science by those in positions of authority and responsibility as well as by those who shape opinion is therefore of importance for the national welfare.[5]

How much knowledge and familiarity with science is desirable? Or, to put it realistically, how much is absolutely necessary? Other than having a thoroughly trained scientist in a position of administrative power—and what is more, a different specialist each time the problem changes—there is no alternative to having a president or cabinet member or department head who cannot be expected to possess complete understanding of the science involved in major decisions such as those which have had to be made during the last two or more decades. Where a decision involving many related factors has to be made with reference to one of them, however, it cannot be made without taking into account how it will affect all the rest. Only the person responsible for the welfare of the whole structure, therefore, can have the right to make the final decision relating to any one part, especially as an expert on only one of the matters will be more likely to see the claims of that to the exclusion of the rest than one who is less specialized in his relation to it. But even if we could assume a perfectly ideal situation in which on any given project scientific opinion was unanimous and only a clear decision lay before an administrator who was scientifically knowledgeable, there would still remain the question of which scientific projects were to receive priority, how much support they were to receive, how they would affect plans for all the other multifarious activities of a state, and so on. The decision would ultimately have to be related to large national goals, immediate over-all policy, and economics. It would, that is, be, in the final analysis, a non-scientific decision. "Scientific discoveries," writes Don Price, "do not restrict the scope of political and administrative discretion any more than they reduce the possibilities of further scientific research. On the contrary, they enlarge the opportunities and broaden the possibilities for discretionary judgment in governmental affairs just as they do for the acquisition of further knowledge." [6] What further complicates the life of a public servant in reviewing scientific problems is that almost invariably there is some division of opinion among experts. The danger comes not only from the ignorance of the administrator who is a non-expert—a danger which is always easy to exaggerate on the part of the expert—but from the division among the experts.

Recent history offers us some interesting cases of this sort. President Truman made the decision to approve the project which produced the fusion, or hydrogen, bomb in the face of strongly divided opinion among distinguished scientists who apparently could not agree on either the feasibility of such a weapon or the desirability of giving it priority over others. This conspicuously non-scientific man was able to make this decision under difficult circumstances because it involved his judgment as to how probable a successful outcome might be in view of the strong conviction of some distinguished experts, and how much of a gamble the circumstances allowed him to take on this one possibility. The decision, in short, was a common-sense decision and did not involve him as a scientist. As things turned out, it was, scientifically at least, a correct gamble. Was it the right decision other than scientifically? This question still arouses differences of opinion among scientists, politicians, military men, and the interested public. Could Truman have arrived at a decision more expeditiously or correctly if he had had a greater degree of scientific knowledge? He could never be expected to know as much nuclear physics as the scientists who disagreed with one another, yet he might have known just enough to incline him toward one group rather than the other on scientific grounds—not necessarily a better state of affairs.

All decisions of such magnitude have political implications. Men whose careers have been political cannot avoid having political reflexes, yet sound advice can be obtained only if the men selected are scientists of the first order and detached enough from extraneous issues to be impartial. Can political leaders be trusted always to select men on this basis? Eugene Rabinowitch complains that political leaders often "pick out, among dissenting scientific opinions, the ones which fit best their political plans, and not the ones which carry the best scientific support." [7] There is no doubt that this danger exists, but it is not unreasonable to ask, how can political leaders know which ones carry the best scientific support? Only a scientist would know, but which scientist? Would the matter be decided by a majority vote of scientists? by seniority? This is not how scientific matters are decided. Selection of politically favorable advisers is not the only danger. There is the possibility that a man in political life will take the wrong advice because he has placed his trust unreasoningly in a bad personal adviser. This is the case C. P. Snow tries to make out against Churchill and his reliance on Lindemann, the old problem of the monarch's favorite in its modern form.[8]

At bottom, however, the really inescapable difficulty in all cases arises from the differences of opinion among persons who are qualified and who are expected to know—the experts, the final court of appeal. No statesman will ever know enough science to decide a controversy about a scientific matter on the basis of its scientific merits; only another scientist can do so, and he will at once become a party to the controversy. The recent attempt to encircle upper space with a band of fine copper needles is said to have had the full approval of the president's scientific advisers, but from the dissenting voices which followed the experiment it is clear that some important American scientists were not consulted, not to mention some distinguished and very articulate British astronomers. The impartiality of scientists and their willingness to submit to evidence and proof are celebrated. These traditional virtues of the scholar they have elevated to the status of a method. But scientists do have differences of opinion and their judgment can be influenced by their political and other biases. There are indications that in the disputes of the last two decades the stand which individuals took with reference to matters of policy in the development of nuclear power was not unrelated to whether they were pacifists, opponents of communism, sympathetic to communism, old-line conservatives, or liberals. For the really vital areas of decision have not been entirely scientific. Men who disagreed with each other with feeling during some of the memorable episodes in the disputes over policy in the development of nuclear power have nevertheless still been able to read each other's technical articles with approval. Detachment and impartiality do not apparently transfer without loss to non-scientific areas. The absolute virtues of science are effective only where the matters at issue are absolutely scientific.

An interesting case in point has confronted the public in the first public debate over the cessation of the testing of nuclear explosives. The issue on which much of the discussion hinged was the danger from fallout. There was great public interest in the debate, and scientists of great eminence took part—men who have received international recognition for their scientific achievements. Everything that could be known about radioactivity and fallout was known by those scientists who took their views to the public. The various aspects of the issue were fully and seriously discussed, and all the major forms of public communication gave extensive and dignified coverage. And yet disagreement could hardly have been more complete. Many non-scientists who followed this discussion seriously and anixously found themselves in a state of helpless indecision, yet their plight would not have been different if they had had a better scientific education. However good, it could certainly not have been better than that of those who opposed each other in the debate and who labored to influence public opinion. The reason for the difficulty can be found in the nature of the argument. The established facts about fallout were known to both sides, and the dangers of radiation were agreed upon. The difference was only whether the dangers were presumed to be great or slight, and whether they were presumed to be necessary or unnecessary. The question of the degree of the danger, while a scientific question, proved difficult to decide, both because of lack of agreement on what constitutes too much exposure to radiation and because of insufficient evidence about the long-term effect. In consequence the argument was shifted onto issues that bore only indirectly on the scientific question. For example, one contention of those who favored continuation of the tests was that, though a risk was involved, it was necessary, and that the acceptance of such risks was not unusual. The argument was that the possible increase in the hazards of radiation, such as incidence of bone cancer or the likelihood of undesirable mutations in the future, must be accepted in the interest of scientific progress as well as national strength, just as we accept other risks, or as we accept the inevitability of deaths from traffic accidents in order to have the convenience of automobiles. This is a scientific-seeming proposition.

It was made by a scientist, it suggests a statistical element, it appeals to our faith in technological progress, and it is future-directed. Nevertheless, it is not a scientific argument and the decision it supports is not a scientific decision. It is a moral decision, and the argument is a moral argument. The reasoning is by analogy, long known to logicians as a useful form of argument for which there is no adequate method of formal verification. The analogy used is, moreover, imperfect. Asking a society to take collective risks when it is in a position to accept or reject them is not the same as asking it to approve risks which may affect the welfare of those who are unable to participate in the decision but who will have to bear the penalty of the risk's having been taken. The attitude is future-directed, but the notion of the future is determined by considerations primarily technical and scientific; it gambles the chances of a future scientific good against the chances of a possible future human ill. This particular argument received considerable criticism at the time, but the point in discussing it is not to prove that the other side had greater merit. It serves to illustrate the kind of argument which was used by men of science, and in this respect the arguments on the other side were not different in kind once the scientific matters were disposed of. At bottom, the main issues were not scientific; they were moral and political, and the present state of the argument has not altered their character.

This is not an apology for scientific ignorance; but it is completely unrealistic to suppose that improving the scientific education of men in public life and government will alter in any basic way the difficulties involved in making important decisions relating to military strength, diplomacy, or public welfare in those areas where government and science meet. In all such cases there will be some issues of a purely scientific nature on which scientists will be in substantial agreement. There will secondly enter into the case some elements which are scientifically gray, upon which even scientists cannot be certain and hence about which they can disagree. And finally there will also enter considerations which call for common sense, respect for the claims of all the factors bearing on the national interest, and administrative skill, rather than for scientific knowledge or understanding. The first clearly belongs to the scientist; no layman can ever be accomplished enough to qualify, and where scientific opinion is undivided only a megalomaniac or madman would disregard today the scientific concensus. The second also belongs to the scientist, but here even the scientist is likely to find himself on ambiguous and dangerous ground where knowledge is uncertain, choices exist, and his impartiality is not inviolate. The important service which scientific advisers can perform at this point is not only to make the scientific certainties clear but also to indicate what remains in a state of scientific uncertainty and why, and so to place into relief the elements of the question that lie outside the province of scientific knowledge and opinion. Outside the area of scientific certainty and unanimity, however, few guides to conduct or rules of thumb can be laid down, and the wise administrator would adopt an attitude of skepticism toward these in any case. One can only hope that the final decision would rest with some one whose intelligence was sufficiently cultivated to listen with understanding and discrimination to what scientists had to tell him, and who possessed common sense, judgment, and the political ability to see that what it is right to do gets done. These attributes will be more valuable than an elaborate but still amateur knowledge of science gained perhaps at the expense of the development of his native talents.

Increasingly, as scientists find themselves in continuous close touch with government officials, they find themselves called upon to render services which are not simply those of a qualified adviser. The question has therefore arisen, in what general ways does his scientific training affect a scientist's fitness for wide administrative responsibilities in the governments of today? There was a time not too long ago when one heard that, if those who conducted the affairs of the world had been trained in the methods of science and in scientific impartiality, things in the practical world would not always be in the characteristic mess in which one usually found them. This opinion is less often heard today. Scientists have become more sophisticated about science and its methods, and the case could not be more straightforwardly put than it is by Conant:

> To say that all impartial and accurate analyses of facts are examples of scientific method is to add confusion beyond measure to the problems of understanding science. To claim that the study of science is the best education for young men who aspire to become impartial analysts of human affairs is to put forward a very dubious educational hypothesis at best. Indeed those who contend that the habits of thought and the point of view of the scientist as scientist can be transferred with advantage to other human activities have hard work documenting their proposition. Only an occasional brave man will be found nowadays to claim that the so-called scientific method is applicable to the solution of almost all the problems of daily life in the modern world. Yet some proponents of this doctrine have at times gone even further and maintained that only by a widespread application of the scientific method to the problems of society at every level can we hope for peace and sanity.[9]

It seems reasonable to suppose, however, that so thorough a discipline as is the training for science would develop attitudes and habits peculiar to men of science and perhaps of special usefulness today in the world of affairs. The most recent, and in some ways the most original, commentary on this problem is C. P. Snow's *Science and Government*, delivered as the Godkin Lectures at Harvard in 1960. Like all of Snow's recent pronouncements, it brings to bear on this important question his experience in science and administration, and it has created widespread interest. The main problem to which Snow addresses himself is the decisions—many having the most fateful implications—which are administratively made under conditions that avoid both the free debate of legislators and public officials and also the free discussion of scientific men, in short, secret science. The discussion is initiated by an account of an extraordinary episode in England during the last war—Tizard's great services in making the decision and directing the project which gave England an effective radar and thus won the Battle of Britain, and the replacement of Tizard with Lindemann after Churchill came to power. It is a fascinating story, so effectively told that it reads like a novel. It reads, in fact, like a novel by C. P. Snow with its interest in the nature of struggles for power and in the character of people who become involved in them.[10] This story becomes a paradigm or exemplum from which Snow derives some lessons for future management of governmental decisions involving science. The story and its lessons provide the setting for some further observations on the qualifications of scientists for public affairs today. Though there is no substitute for following Snow's argument in all its completeness, the views which have a bearing on the present discussion can be considered independently, on their own merits.

The special virtue which Snow sees in the scientist today is that in his attitude and habits of mind he is less likely to be touched with the prevailing philosophic disease of our times that handicaps Western Europe and the United States in their competition with more dynamic societies:

> We are becoming existential societies—and we are living in the same world with future-directed societies. This existential flavour is obvious in our art. In fact, we are becoming unable to accept any other kind of art. . . . We seem to be flexible, but we haven't any model of the future before us. In the significant sense, we can't change. And to change is what we have to do. *That* is why I want scientists active in all levels of government. . . . I make a special requirement for the scientists proper, because, partly by training, partly by self-selection, they include a number of speculative and socially imaginative minds. . . . I believe scientists have something to give which our kind of existential society is desperately short of: so short of, that it fails to recognise of what it is starved. That is foresight.[11]

Snow does not contend that foresight is a quality which all scientists pick up in their training and research as they pick up the habit of impartiality, but rather that those who already possess this rare gift are made better in it by virtue of their practice of science:

> I suppose most scientists possess nothing of this foresight. But if they have any trace of the capability, then their experience, more than any experience at present open to us, gives them the chance to bring it out. For science, by its very nature, exists in history. Any scientist realises that his subject is moving in time—that he knows incomparably more than better, cleverer, and deeper men did twenty years ago. He knows that his pupils, in twenty years, will know incomparably more than he does. Scientists have it in them to know what a future-directed society feels like, for science itself, in its human aspect, is just that.[12]

No other activity in Western society, Snow contends, encourages through its training and basic attitudes these extremely valuable qualities. Administrators become "masters of the short-term solution"; artists, writers, and generally educated men in our society are inclined toward an existentially pessimistic or indifferent view of the future. The refreshing virtue which the scientist can bring to the conduct of public affairs today is that his work as a scientist has sharpened the attribute of foresight if he happens to have it and has converted it into a habit and a trait of character. This is the substance of Snow's case. It offers some hope of escape from the disadvantages of a widely recognized disability of Western countries in the present situation—that while they seem to lack a sense of purpose or deep conviction about their destiny they are opposed by countries which are convinced that the future is theirs, and which are sure that they know its form because they are confident of their ability to shape it.

The heart of the problem is in Snow's observation that, though we must perforce change somehow to survive, we cannot change because "we haven't any model for the future." Now the kind of model which would serve the purpose of a confident, "future-directed" society is an idea of what kind of society it wishes for its people, what kinds of separate goals this calls for, and what kind of a state is required to secure them. The problem for its leaders is to determine what means—economic, military, educational, scientific, political, diplomatic—will be needed to bring about the kind of society envisioned. Old, established, flourishing societies retain the idea of their society in their traditions, institutions, and forms; but where the society becomes uncertain of its future prospects or begins to doubt the viability of its traditions and institutions, the decisions of its leaders cannot go much beyond meeting short-term objectives thrust upon the society by contingencies which arise or by threats to its immediate well-being. Foresight in such a case can go no further than judgment on the possibilities open for action and the possible consequences of action applied to the unavoidable pressing demands of the moment; foresight becomes essentially brilliance in tactics, like that of a fine general forced into a defensive war. Assuming that the view of Western society presented by Snow and shared by many others is correct, what can we expect of the scientist's foresight?

One illustration Snow gives of scientific foresight is the memorandum of James Franck and his fellow scientists on the consequences of the atomic bomb. Secretary Stimson's memorandum to President Truman on this same subject had appeared several weeks earlier, a notable case of foresight as Snow concedes, but Stimson's, he believes, shows up less well by comparison. Stimson "had to rely on his political sense," whereas the position of the scientists was characterized by training and knowledge, and something which Snow expresses as "an expectation of knowledge to come." [13] Stimson's foresight might seem to be the more impressive of the two simply by virtue of his ignorance of science, but one must not minimize the significance of the scientists' memorandum, for it was a striking departure from the traditional indifference of scientists to the consequences of their activity. The magnitude of the achievement in producing nuclear fission and creating the bomb impressed many intelligent scientists with the gravity of its consequences, and an encouraging outcome of their concern was their serious inclination to probe the possible consequences of scientific knowledge. This is a valuable extension of the scientist's usual technical preoccupation with the possible consequences of a given hypothesis, or experiment, or new data, and one that society should find useful; but the question is whether the quality revealed in the Franck memorandum represents something more than the statesman's short-range judgment on the possible outcome of policy or action except in being on a technical subject and having the virtue of being informed. Unless the foresight which the scientist brings to the present scene implies also a "model for the future" in any sense other than a prediction of the kind of science and technology we will have to live with, it cannot, by Snow's analysis of our plight, provide the incentive to change in the significant sense which he believes we have to. The foresight of the scientist will provide us with a picture of the changes which must occur in the light of his discoveries and their development, and give us the context within which action must be taken, but this is a limited order of foresight and is not the same as a model of the future in the large sense in which a view of society becomes the guide for decisions and provides a center for collective aspirations and inspires the collective will.

Snow does not go so far as to say that the foresight of the scientist includes this larger view, but there is some indication that he thinks it might. In explaining why he wants scientists at all levels of government, he remarks that "partly by training, partly by self-selection, they include a number of speculative and socially imaginative minds." The word, "socially," jumps out of that sentence like some totally unexpected formation in a familiar landscape. Speculative, imaginative—these are attributes of all good scientists, but the modifier, "socially," is difficult to justify. It does not seem distinctively appropriate when applied to the lives of great scientists or to the character of the average scientist compared to his fellows in other fields. Snow furthermore regards scientific activity as constituting a society with a progressive and forward-looking attitude: "scientists have it within them to know what a future-directed society feels like, for science itself, in its human aspect, is just that." To speak of science as a future-directed society "in its human aspect" is to use the word, "society," in a very special sense Scientists are a society only by virtue of the fact that they are all engaged in the same professional activity, and are bound together by this common interest irrespective of whatever other more complex and binding ties they may have. It is also in a very special sense, too, that this "society" is future-directed. Science as a discipline is progressive and dynamic, since it increases in knowledge and in some respects in complexity, so that the science of the past seems modest in comparison with that of the present, and the science of the present seems timid in comparison with the prospects and possibilities for the future. Continual association with such an activity can encourage a disposition to look with a buoyant spirit toward the future, at least of science, but whether this attitude will be readily transferred over to the somewhat more confused world outside the laboratory, or whether the future-directedness of science will rub off on others with whom the scientist may associate in public life, is quite another matter.

The point is important, since in using the phrase, "future-directed," Snow is by implication equating the attitude of the scientist with that of nations which, unlike those of the West, seem to be confidently turned toward the future. Snow has been particularly effective in presenting certain aspects of the critical position of Western society, and in this connection, he alludes on several occasions to the scientific and industrial dynamism of Russia and the realism of its educational programs. The dynamism of Russia has several aspects. It displays the energy of a revolutionary regime which is determined to eliminate the last traces of a culture that retained the marks of an outmoded society long after its evils were apparent to everyone, and which has the resources and determination to catch up with the most advanced industrial and technological societies in a relatively short time. These are concrete aims, but the energy and future-directedness of Russia come also from the fact that its concrete goals are believed to be essential steps in the formation of a new state and a new society of the future that has become the symbol of what amounts to a state religion. Commenting on the efforts which need to be made to aid underdeveloped countries, Snow notes that in the matter of providing for an adequate number of scientists and engineers for this purpose "the Russians have a clear edge." [14] If the Russians are willing to train scientists and engineers on a prodigal scale, if they are able to assign them to other countries in increasing numbers as we presumably are not, it is in something of the same spirit that the Church once trained its clergy and sent missionaries to remote unchartered portions of the globe to convert the heathen and bring their lands eventually under the Christian sphere of influence. The dynamism and future-directedness of the countries which threaten the West are of a different character from those qualities which scientific activity can induce in the scientist. If the same words can be used to describe both, it is with a different appropriateness in each case.

Scientists, nevertheless, have an important role to play in public affairs, and this need not be limited simply to that of adviser to political leaders on scientific matters. Some who, under the pressure of great national need, left their laboratories to become involved in public affairs, have conducted themselves creditably and even with distinction. It would not be surprising to find more of them in public life as time goes on and exercising a greater degree of responsibility. If this happens, it will not be because of a special virtue in their scientific training. In some respects, the training in science is not ideal for the conduct of affairs. The world which occupies the attention of the scientist is a less contingent world than that of human activity, the means he uses to control it are more rigorous and reliable than statesmen can depend on, and the events with which he deals are neither good nor evil. There are scientists who exemplify all the virtues of their training to a high degree but whose elevation to positions of administrative responsibility would fill their scientific colleagues with horror. What makes it likely that science will produce men capable of leadership in public life is simply that the commanding position of science in the modern world will inevitably attract ambitious men of unusual abilities and great energies. Science is now at the very center of excitement in our culture; it offers all the attractive rewards except that of great wealth; and it is open to men of talent irrespective of their social and economic status and with fewer financial and social barriers than some other desirable professions. Many capable men who are now attracted to science possess what might be called undifferentiated talents —they are not, that is, men with an unusual knack for mathematics, or extraordinary acuity of the senses, or great powers of abstraction. They are simply very able men who can master difficult things if the incentive is sufficient, but who under other circumstances might have been tempted to do something else. These men will make good scientists, but among them will be found individuals who also understand the ways of the contingent world of affairs, who have a taste for action, and who possess a flair for directing affairs. In a similar way and for similar reasons, the Church once offered careers to able men, and some of them became powerful, like Wolsey and Richelieu. At a later date commerce and industry offered unusual opportunities for able and ambitious men, and some of them rose to greatness and sat down with statemen, and, trained in a more ruthless school than the scientist, bought legislators and influenced policy in a less than impartial spirit. But commerce and industry did not appeal strongly to men who, in addition to restless general abilities, had a taste for the intellectual life. Science does. It is therefore likely to produce men who, if they have the gifts which qualify them for participation in the practical world of affairs, will bring to them a rare combination of talents. It is inevitable that the qualities and the character of a man of this type will bear the mark of the training and habits he acquired from his specialized activities. It might make him impatient of sloppy thinking, of evasion of facts, of bias and dishonesty; perhaps, if we want to be speculative, he may be less subject to continual misgivings about the future. It is not possible to predict the precise way in which a man's education will influence the particular set of his energies and the style of his actions. That it will be valuable can be taken for granted. But the qualities which will be responsible for the success of a scientist in public affairs will be those which he has in common with all other men of whatever training who demonstrate a capacity to move in the public world with effectiveness. Somehow, these qualities will have had to find nourishment outside the discipline of science.[15]

In his study of government and science, Don Price pays a tribute to the long and beneficial influence which science has had in American life:

The skeptical and questioning approach of science has played a part in freeing the United States from the authority of old tradition and protecting her from the fanaticism of new ideologies. The restless energy of the scientist and the engineer has broken through the constraints of red tape and supplied a dynamic drive to the development of government programs, as well as to the productiveness of private industry.[16]

In his detailed review of the rapid growth of governmental involvement in science and the administrative complexities which have arisen in consequence, he gives numerous illustrations of disinterested and intelligent effort on the part of scientists who have helped to overcome the difficulties in the way of effective association between scientists and government and who at the same time have striven to preserve the independence and vitality of the nation's scientific activities. As one of the best informed and judicious commentators on this matter, he may perhaps be allowed the last word:

An administrative system is only a reflection of the hopes and beliefs and skills of individual human beings. The top administrative service of the nation cannot exist without the support of informed public opinion or without the participation of men whose appreciation of public affairs is broader than that of any specialty. Here, then, is the opportunity for the American university—to educate in the humanities and the social sciences men who have an understanding of the role of the natural sciences in government and society, and to educate natural scientists who can appreciate the problems faced by the politican and the administrator, and who will, some of them, shoulder the burdens of direct administration of national affairs.[17]

V
Science for the Non-Scientist

To the scientifically minded observer, the modern world is made up largely of scientific illiterates muddling their way through a complex scientific civilization. The widespread lack of knowledge and understanding of science is a theme which inspires a wide range of feelings from impatience to sober anxiety over our lack of readiness for the scientific age in which we live. Our general scientific ignorance is regarded as a serious danger, responsible for our failure to enlist and train an adequate number of scientists and engineers, and at the basis of the separation in our culture that divides most of the educated public and our intellectual and political leaders from the distinctive creative activity of our times.

Since the failure is one of education, the remedy is usually sought in some form of change in method and emphasis in college and university courses of study. Bentley Glass, for example, would organize a liberal education around science:

> The major problem of higher education is to cure this growing schizophrenia. The sciences must become the core of a liberal education . . . , but "in teaching science we must not forget . . . that it is simultaneously social study and creative art, a history of ideas, a philosophy, and a supreme product of esthetic ingenuity." The humanities and social sciences, on their part, require more than a cognizance that the natural sciences exist. They must become permeated with the knowledge of science if they are to be more than a relic of a buried age.[1]

Though the same writer elsewhere makes clear that the core

is not the whole apple, once the notion of the preeminence of science becomes the guiding principle of education, it tends to subordinate the other disciplines and assign them a role of handmaidens of science. In this spirit, it has been suggested by Joseph Gallant that one role of the humanities could be to promote the cause of science:

> Where, then, can the emotional and imaginative appeal of modern science be conveyed to the student? Obviously in the humanities and the arts, generally, and most specifically, in the study of literature. . . . It is literature which can most appropriately project the emotional impact of the scientific outlook of the individual.

The humanities, this writer continues, can incline the scientifically minded to a career in science "by showing them the wonder, the vision, and the excitement of science," and they can inspire others "to share in the scientific outlook and, through it, to enrich the arts and the literature of the future." [2] In essence, such proposals amount to a primary concern for the needs of science and the welfare of scientific activity, and they solve the problem of the division in our culture by identifying the welfare of all learning with the welfare of the scientific. In effect, they undermine the independent validity of the other disciplines. They fail, therefore, to answer the question of what should be the scientific education of persons who are to live in a technologically advanced society in which science is an activity of the highest importance, but who are not going to devote their lives to science and who have sufficient respect for the social relevance of their profession and the intellectual relevance of their branch of learning to regard it as something more than a servant of science. To answer this question, it is necessary to accept the proposition that we live in a scientific age without accepting as a necessary corollary that science is the sole important and primary preoccupation of our civilization.

In this form the problem is not new, and for many years it has interested educators who were not scientists. Only in recent years, however, have scientists themselves regarded it with concern—only, that is, since the events of the last two decades have thrust science into prominence and made everything connected with it a matter of great urgency. Looking back at the recent past from their present vantage, it is understandable that scientists should regard with disapproval the weakness of our educational arrangements from the point of view of the education of scientists. It is understandable also that they should view the inferior general scientific knowledge of the non-scientist with disapproval and should regard our efforts to produce an enlightened society for the scientific age as a failure. To the best of my knowledge, no one has seriously suggested that the scientific fraternity might possibly bear a large share of the responsibility for this particular failure. It is a hypothesis worth considering, not for the purpose of identifying the villian of the story, if there is one, but because in the reasons for past failures may be found the clues for future improvement.

For the last forty or so years it has been virtually impossible for a liberal arts student in most colleges and universities in the United States to receive a degree without having taken at least one year-course in science. Most baccalaureate degrees require two year-courses during the first two years. This may not look like a great deal, but it represents approximately one-fourth of the student's total academic time during the first two years. These courses have been taught by scientists, and they have afforded a considerable opportunity, if not to allow for a comprehensive view of the state of scientific knowledge in many fields, at least to provide an insight into the great accomplishments of science in the past and in our time and to instill some enthusiasm for science and its accomplishments and an appreciation of the beauty and magnitude of the scientific achievements of man. Scientists attribute the lack of great success to a want of sufficient time for a comprehensive scientific education, to the poor preparation of the non-scientific students, and to the incapacity of many of them for the discipline of the mathematical aspects of science and the exactness of laboratory work. All these things may be true, but the fact still remains that for the better part of a half a century the teachers of beginning science in colleges and universities have had a fair opportunity to make things come out differently.

They should not be too seriously condemned. During the

same period American science came of age, and the first consideration and primary task of American academic scientists was the training of scientists. The average scientist accepted the responsibility of teaching non-scientists as a task assigned him by his college or university, and if he was devoted to his subject he accepted these students with good-natured indulgence or missionary devotion. But neither his view of his primary responsibility to the training of future scientists nor any sense of urgency about providing for a scientifically enlightened society inclined him to consider the teaching of non-scientist students as a matter of special importance. Besides, he believed that science was science, and that a course which prepared a young student to go on to more advanced work in any particular science was the best introduction to science for everyone. In consequence, most colleges offered a series of one-year introductory courses, chiefly in physics, chemistry, biology, geology, and sometimes astronomy, from which freshman and sophomore students could choose the number of units of science required for a degree. In most schools, these courses combined lectures and laboratory instruction. With some variation, this approach to the teaching of science to non-scientists has dominated the American college campus for something like thirty to fifty years. There have been occasional modifications. Sometimes the students who had some preparation and intended to continue in the science have been placed in a different course, but the course for the rest has not been different in kind but simply "easier." There were also occasionally developed, usually under pressure from reforming deans and presidents, courses in Basic Science or General Science which were more comprehensive in substance and somewhat different in method of instruction; but these have been palliative measures and have represented no fundamentally original approach. Most scientists, chiefly interested in those of their students who wished to continue to advanced courses and even more interested in their own research, were reluctant to undertake a serious philosophic and educational reappraisal of what they were doing for students who would never be scientists and then to devote the amount of exhausting work which would be required to put a new set of courses into practice for a purpose which they could not—at least formerly—take seriously, and which was after all not their primary purpose. It is uncharitable to reproach them for so eminently reasonable an attitude, even though today they regard the scientifically naïve educated man as an anomaly, and even a danger, in this age.

The disadvantage of the prevailing arrangement for the non-scientist student was that he could meet the science requirements for his degree, even when these amounted to one-fourth of his academic time for two years, without acquiring a decent idea of what science, and most important of all, what modern science, was all about. The courses which he might take were, with a few notable exceptions, designed to prepare a student for the next advanced course in the department. The introductory courses could not fail to do the non-scientist some good, but they left him far short of his proper goal. To be sure, a comprehensive grasp of the situation in modern science is beyond even the professional scientists, but because as any given science reaches an advanced level the distinction between special branches often disappears and the knowledge and techniques of a neighboring science becomes necessary, a person who makes a career of science eventually acquires an idea of what is going on in science all around him. Besides, he has the advantage of talking to scientific colleagues on a technically knowledgeable footing, and his horizon is continually widening. The student who will not be a scientist suffers acutely therefore from the curriculum which has been designed for him. He takes at most two courses, usually in two sciences, and then his formal contact with science ceases, and unless he is extraordinarily curious and enterprising, his view of science is circumscribed within this narrow confine. If he wishes to increase his comprehension while in college, he can either take more advanced courses in one of the sciences to which he has been introduced or continue to take introductory courses in other sciences. Neither of these alternatives represents an optimum use of the limited amount of time he has as a student. The first leads to specialization, which he can ill afford, and the other, while preferable, still leaves him far behind in the kind of sophistication of outlook and perception which he should acquire. The conventional system has not, therefore, provided a proper preparation for a lifetime of interest in the subject. The courses have been taught so that the next logical step would be in the direction of increased proficiency in technique and more specialized subject matter, and not toward the gratification of a layman's curiosity and his philosophical and historical interest. They have not conditioned students to carry their interest into their general reading for the rest of their lives. Those nonscientists of our times who have tried to keep alive an interest in the developments in science have simply been the kind of intelligent and curious minds who cannot be prevented from trying to find out what is going on around them. As for the college science courses for beginners, they have usually been regarded as a terminal experience for the non-scientist by both parties.

One unfortunate consequence of the standard scientific fare for the non-scientist student has been that it has not induced an understanding of the nature of the scientific enterprise, nor a very subtle notion of its character. The student hears a great deal of talk about scientific method, as though it were some potent key, difficult of access to the outsider, for unlocking Nature's secrets; but from the way in which he has usually been taught, it is never clear to him whether this means anything more than laboratory technique. It is rarely that a student ever finds out how, if at all, scientific method differs from any other form of realistic approach to evidence and hard, rigorous thinking—how, that is, it might differ from what goes on when a detective determines the nature of the crime and identifies the criminal, or a judge decides whether a defendant is guilty or not guilty, or a literary scholar determines whether a document is a forgery. He is told that science demands accurate observation, but almost never does he come to realize that in most cases scientists cannot directly observe the phenomena they are studying and that most scientific work is a highly specialized form of deducing from circumstantial evidence. The student sometimes gets a glimpse of the way science works from demonstrations of the experimental basis of the important laws and hypotheses of his particular science, but it is surprising how incomplete is his grasp of the reasoning underlying some of the classic discoveries of science.

I became interested many years ago, quite by accident, in the understanding which students possessed of the replacing of the Ptolemaic system by the Copernican, certainly, a significant change and an important episode in the whole exciting story of the establishment of modern science during the sixteenth and seventeenth centuries. I found that most students were aware that the difference between the two theories of the planetary system was that in the Ptolemaic the earth was at the center and that in the Copernican the sun was at the center and the planets, including the earth, circled the sun in regular orbits. A few were also aware vaguely that the Copernican system was not immediately accepted by most of the learned men of the time. Beyond this, the appreciation of the scientific aspects of this revolution was very disappointing. Of the persons who knew this much, only a very small portion also knew that an important step in establishing the Copernican system was Galileo's observations with his improved telescope, but of this small number even fewer could say what it was that Galileo saw—except that he probably did not see the planets wheeling around the sun. One might find a student who could report that what Galileo saw was that there were satellites circling Jupiter and that Venus went through phases like our moon. But I never had the luck to come upon a student, scientist or non-scientist, who knew this much and who could also explain what these observations demonstrated—that the satellites had no bearing as scientific proof of Copernicus' theory but were significant in destroying a prevailing notion based on curious analogical thinking that there could be only seven planets, and that the observation of the phases of Venus was scientifically significant because it introduced a datum that could be accounted for by the Copernican scheme but not by the Ptolemaic. Today the results of a similar probe might produce more satisfactory results, since there is reason to believe that the training of some students is sharper than it once was; but the record of non-scientists would not, I believe, be sufficiently better to eliminate the wonder that so excellent an opportunity on so important a topic could be so effectively lost.

One by-product of the traditional scheme of undergraduate science instruction has been that the non-scientist student has failed to develop a reasonable, intelligent attitude toward scientific activity and toward scientists. Many students leave their limited experience with science without coming to appreciate it as a thoroughly human enterprise which is not beyond the reach of a reasonably intelligent individual willing to undergo the discipline required of any serious intellectual endeavor, and which is worth pursuing outside the classroom. Some twenty years ago there appeared a textbook in physics which attempted to combine the usual basic material with information concerning the steps by which certain scientific concepts became established, the philosophic issues underlying some developments in science, and even the influence of the milieu on the appearance and the development of particular scientific ideas. I found that physicists on the whole did not share my enthusiasm for this text. One objection was that the text was too subtle and complex for students—hard enough to teach Ohm's law without getting involved in the irrelevant business of how the concept of an electric current came into being, or to get across the basic principles of mechanics without introducing the philosophical difficulties which were encountered in establishing the concept of mass. Another objection was that it required the instructor to teach, as one physicist put it, what isn't so any more; that is, there was no point in talking about the wrong inferences which one scientist drew from significant original observations or experiments in order to deal with the correct inferences which another scientist drew from them. These objections are understandable and to some degree valid, but they simply call attention to the limited view of science and scientific activity the beginning student was likely to acquire from the approved methods of elementary science instruction. The science student who goes on to build on his introductory material can acquire a deeper understanding about science as he progresses, but for the student whose experience with a given science is terminal with the elementary course, there is a real danger that the science will take on the appearance of a body of commandments inscribed on tablets brought down from some scientific Mt. Sinai by a succession of prophets.

Difficulties of attitude have, I believe, plagued both sides. Except for the unusual brilliant student, there is a barrier between the non-scientist and the scientist in the average introductory course, a subtle lack of real understanding and approval, which no amount of good will can fully eliminate. What the student contributes to this vague lack of rapport has been often enough gone into. But I believe that the average scientist—unless he has a little of the missionary spirit about his science—contributes something through a lack of real appreciation of the non-scientist's intellectual goals. Anyone who is around scientists and engineers on university faculties discovers that some of them have a way of referring to "easy" or "soft" subjects, by which they mean most of the subjects outside the scientific.[3] Their views are no doubt colored by sad experiences with college students who do not have a natural taste for science and who are ill prepared for it, and who are often too stupid for it.

But this feeling arises from other, deeper sources. Scientists perceive that they can understand the fruits of serious investigations in many other disciplines in a way which the scholars in them cannot understand science. A scientist who reads a new piece of historical writing with understanding and even critical appraisal may well conclude that history requires a somewhat lower order of powers than science, since historians are helpless in the face of scientific research at a level comparable to their own in history. Scientists are prone to attribute these differences to the exactness and intellectual rigor demanded by science as much as to the specialized conceptual and symbolic schemes employed by science. It must be admitted that fuzziness and sloppiness have less chance of surviving in science than in some other branches of learning, and it is less easy to get by with fraud or a plausible piece of pretentious nonsense. The beauty of its rigor is one of the fine things which exposure to the scientific discipline has to offer. But scientists have been somewhat too ready to attribute wholly to the scientist the virtues which are the property of their science. The capacity of the trained scientist to solve problems routinely with an elegance and finality which is not shared by scholars in other fields is in large part due to the fact that by its nature the scientific problem lends itself more readily to definitive solution, and the elaborate system of scientific activity which now exists contributes greatly to maintaining this activity at a competent level and protecting its practitioners and itself against disaster and fraud. On this theme Conant writes:

> Would it be too much to say that in the natural sciences today the given social environment has made it very easy for even an emotionally unstable person to be exact and impartial in his laboratory. The tradition he inherits, his instruments, the high degree of specialization, the crowd of witnesses that surrounds him, so to speak (if he publishes his results)—these all exert pressures that make impartiality on matters of *his* science almost automatic. . . . Once science became a self-propagating social phenomenon, those who till these fields have had a relatively easy time keeping up with the tradition of their forebears.[4]

The non-scientist almost never appreciates this fact, and he is easily intimidated. Many scientists, however, do not appreciate it either, and the view which they give of science has had the effect of adding in a subtle way to all the other forces which encourage alienation of the outsider.

Scientists are sometimes wont to deplore the naïve attitude of the public toward science, and complain to the effect that "science is respected, but, perhaps, among non-scientists (and this includes teachers and writers), with the same mixed fear and regard felt for the medicine man among primitive groups."[5] There is a small modicum of truth in this extravagant statement, but if the case is really as it is represented, there is little wonder. With some notable exceptions, most of the introduction to science which a non-scientist student gets in college leaves him with little grounds for genuine understanding and less for a sophistication of attitude. I believe, too, that so far as the more extraordinary aspects of modern science go, too many scientists have enjoyed surprising and mystifying the outsider. And as for the great scientists, in such historical information as the student is likely to have given him, they are represented to him as figures in a pantheon, each individual associated with one or two celebrated discoveries which bear his name, who seem not to have made a mess of their lives or sullied themselves with scandal like artists, poets, and musicians. A little less awe, as David Hawkins suggests, would do no harm:

> On the whole, the role of science in the usual story of modern culture is rather stilted. Science is there in every scene, performing some momentous feat of discovery. It is treated very respectfully, and sometimes with awe and reverence, or at least with the kind of enthusiasm that may pass for awe and reverence. There are some critics who do not share this enthusiasm; they excite a good deal of sympathy sometimes, but of course they are quite wrong. Actually the critics of science have this important virtue, that they make it come alive. The official eulogies, on the other hand, accord science the treatment usually reserved for famous admirals and ex-presidents in the movies. It is better to be presented as a villain, in some part of your true character, or even to be misrepresented as a lively and believable rascal, than to appear as a plaster saint or wooden Indian or a Man from Mars in a space-helmet.[6]

The tradition of teaching science to the non-scientist in what might be called The Authorized Version has resulted in a partly educated public, too naïve, in the judgment of scientists, to cope with the phenomenon of science today. This is not, as the physicist Polycarp Kusch implies in the following statement, to the advantage of science:

> We have impaired the ability of those not trained in science to understand it by too great an emphasis on the power of science without an occasional digression on its limitations . . . The lack of a sharp awareness of the limitations of science may be downright dangerous for the layman. Bombarded as he is with news of the triumphs of science, he may feel that science will solve all problems. He may believe that science can produce any miracle necessary to solve a problem, and he often attributes to science a quality of wisdom that is wholly outside its sphere.[7]

There would be less likelihood of such dangerous naïveté if

the scientific education for the layman introduced him to a proper understanding of science and scientific activity and gave him the basis and incentive for continued, revivifying interest in its accomplishments.

If we are going to provide in the future a better scientific education for the non-scientist, there should be some agreement as to the purpose of such an education. The aim of all education which is not designed to train an individual for a profession is to make him at home in the world in which he lives, or more accurately in the culture in which he lives. What is called for by this aim does not remain constant. Today it is less vital than it once was to be learned in the doctrine of transubstantiation or prevenient grace. Today science is an activity of prime importance and enters everywhere in some form or another into our lives. As the scientist is likely to see it, no other intellectual activity of our day has such claims on our minds and our sensibilities:

> The aspect of Western Society which has differentiated it from the rest of the world in the last four centuries has been its concern with science. Science is the bed rock of the contemporary world. Without the Weltanschauung of modern science, no form of thinking, feeling, or reacting has validity today. No man can see the world except through his modern eyes, and these, in a large measure, are conditioned by the scientific outlook, whether or not he is conscious of it. Not to apprehend this world from the standpoint of science is, therefore, to belie the very process of seeing. To speak in any idiom other than that which incorporates the scientific outlook is to speak the language of the dead—a feat which usually falsifies the meanings and the nuances which that language had for those who lived in the past.[8]

If there is any argument with this statement it is with its tendency to exaggeration and with the trace of belligerence in the rhetoric. But it is one thing to accept in general the view which is expressed and another to determine where it leads us. It led the writer of it to recommend that teachers of literature undertake the role of inspiring students to share in the scientific outlook and to choose science for a career.

The instruction we provide for students will be in part influenced by the specific objectives we have for it.

One view of the purpose of scientific education is that it
will help us to adjust to the extraordinary technological devices which surround us in our daily lives. The housewife's
incomprehension of the workings of an electric cord, the
driver's lack of understanding of internal combustion en-
gines, the passenger's indifference to the aerodynamics and
chemistry of his jet plane—such forms of ignorance create a
barrier, it is alleged, between us and the products of our culture, which makes us apprehensive of our environment and
at the least deprives us of pleasure in the everyday technology
which sustains our society. There is, to be sure, some virtue
in all knowledge, and those who do know what underlies
the operation of the fabulous contrivances which surround
us have that advantage over those who do not. It is a misap-
prehension, however, to suppose that people feel more at
home among the machines and devices of our civilization
if they have a substantial knowledge of the science and engineering which underlies the operation of our technical mar-
vels. It is surprising how much at home most people in Western countries, and especially in the United States, do feel in
the twentieth-century technological milieu while remaining
in ignorance of what goes on inside. Men and women in
America drive automobiles with an assurance, a sense of control and participation, and a feeling of empathy, without
really understanding the ignition system or automatic transmission. They are attuned to their machines, feel at home in
them, and are responsive to their sounds with a sixth sense of
when these mean trouble—like a mother who senses danger
signals in the way her child shouts at play. We take delight
in the flight of a well-designed plane, and even have a crude
understanding of the thrust that keeps the plane aloft without knowing how the fuel is translated into that thrust. A
modern woman manipulates the controls on her automatic
washer, adapting its cycles to her wishes in the manner of an
organist arranging the stops on his organ.

It is not necessary really to know more. There is at one's
beck an elaborate system of service and rescue operations
with special equipment and skills which it would be as use-

less as unnecessary to master. That we have acquired an empathy and sense of mastery over our mechanical environment is a tribute to the success of our technology, and the greater its success the more compatible we feel without a compensating increase in knowledge. Until we had arrived at this point of perfection and specialization, the number who could participate in the operation of the products of our technology was limited and our uncertainty and apprehension about them was greater, as in the early days of the automobile when it was necessary to carry along tools, spare parts, and someone who knew how to use them. Today young people grow up with the automobile in a relationship that is not so much that of driver to car as of the two parts of a centaur. It is familiarity rather than competent technical knowledge which develops expertness of use and reduces fear. Failure to appreciate the nature of the layman's relation to our technology leads at times to improper deductions about his attitudes and conduct. Thus it is wrong, for instance, to say that "because we know how gunpowder works, we sigh for the days before atomic bombs." [9] Most people are as ignorant of the chemistry and physics of gunpowder as of the atomic bomb. If they quite understandably sigh for the days of gunpowder it is because they do, in a very important sense, know how atomic bombs work. The purpose of studying science is not primarily to feel at home in our technological jungle. This view would give an incorrect bias to the teaching of non-scientists. It has already been largely responsible for the emphasis on applied rather than theoretical aspects of chemistry and physics in textbooks for secondary schools.

There may be more justification for the view that the general purpose of scientific instruction is to create an informed public capable of taking intelligent part in those decisions of our society which involve science and scientific activity. Conant states the case persuasively:

> The intelligent citizen of today has need to understand both science and scientists. Government officials, businessmen, and trustees of hospitals are often confronted with technical problems; even the "man on the street" must be prepared to pass judgment on new scientific enterprises.

As a voter and as a potential contributor to such agencies as the American Cancer Society, he or she may well ask searching questions about the organization of scientific endeavors. Does medical research require more or less organization? Is the freedom of the scientist endangered by the vast expenditure of federal funds? Are the universities, the research institutes, government bureaus, or industry the proper place for the scientific pioneers of this century? [10]

These questions, and others like them, require some degree of understanding of the scientific activity, but they are not scientific questions. If we keep this objective too seriously before us in shaping courses in science for the undergraduate we may give them an ineffective direction. To take the concrete case mentioned above, questions which we might ask as contributors to cancer research could be better answered by a comprehensive, unbiased, detailed report on the whole phenomenon of cancer research and of the way the money is used and distributed—a much more difficult sort of information to come by at the moment than readable reports of a scientific nature on various aspects of cancer and cancer research. Like the adaptation to our technological environment, the cultivation of enlightened wisdom in public affairs involving science may require less knowledge of science than is usually supposed, and in any event could best be achieved as an almost unsought-for consequence of a properly conceived and properly conducted scientific education for those who will not follow a scientific career.

The primary aims of such a scientific education for college students should be thought of in intellectual terms. If this education is adequate and successful, the secondary aims will be achieved as a by-product. It should provide knowledge and understanding of science and form the basis and incentive for future continued interest and study. The information the student acquires of the contributions of science to knowledge should include what might be called the classics of scientific discovery—those achievements (usually a series of achievements) which provided important insights into the nature of the physical world and established important new concepts. And it should include some information concerning the present state of scientific knowledge—its most important achievements, the areas of principal interest, and the relation of these achievements to those of the past. The student should further acquire an understanding of the distinctive methods of science, and this should include some notion of the kinds of questions which science can properly be expected to answer and the limits within which its methods are likely to prove successful. Finally, he should acquire some perception of the way in which science is related to its environment—how it is influenced by and in turn influences the thought and activity of the society in which it flourishes. It will be thought that this is too much—that it is, in fact, impossible; and in one sense it is, just as the mastery of the subject matter of history and of historiography is impossible, or again of world literature and critical theory and method. It certainly is chimerically impossible in terms of what has been regarded as standard scientific instruction during the past. In those terms, few scientists on a university faculty could qualify, and what is not expected of a scientist cannot be expected of one who is not. The gap between science and other learning is very great and is daily increasing. It cannot be reduced by anything short of a complete revision of the approach to the teaching of science to non-scientists.

One of the stumbling blocks to a new approach to the teaching of science has been a slight but significant misunderstanding of the nature of science. From the point of view of the practicing scientist, science is a creative activity, and in this respect is related to the arts—to music, or painting, or poetry. What separates it, among other things, from the fine arts and literature is that in these the perceptions of the imagination and insight are supported and rendered valid by means inherent in the particular art and do not need to be validated by the criteria of scholarship. Those in science do. The methods of science are a highly specialized variant of the disciplines and controls of scholarship, developed for the particular needs of scientific activity. A further distinction is that the products of science do not necessarily end in themselves but may provide the basis for further and different creative activity in engineering and other applied fields. The dual aspect of science as an independent creative activity and as a form of scholarship, and beyond this as a power behind technology, has confused the problem of its presentation to students who do not wish to qualify as practitioners. What the scientists have been saying is that to learn anything about science one must do science as a scientist or prospective scientist must do it. Otherwise we would not be teaching science; we would be talking *about* science, which is not the same thing and will not do at all. What the scientists have for the most part been asking for science, however, they would not ask for music, or literature, or art. Scientists have frequently alluded to the unusual interest shown by scientists in music and their competence as performers. Yet of the scientists who are appreciative of music and can claim with justice to understand and enjoy it, very few have studied any appreciable amount of harmony or counterpoint, written music in various styles, studied the technical aspects of the modern systems of avoiding tonality, and composed in the twelve tone row. Yet, *mutatis mutandis,* this is what in effect they ask of the non-scientist in science. It is generally recognized that some practice in the arts and letters is in itself rewarding and a great advantage in their study and appreciation; but all university instruction in these areas assumes that understanding about them which is scholarly, useful, and appreciative is possible without the need of acquiring proficiency in them.

Unless a similar approach to science can be adopted, the case for the non-scientist in our society is virtually hopeless. We cannot close the enormous gap between science and those who do not practice science by providing sufficient conventional instruction in science to make the non-scientist proficient in science. The only hope lies in trying to introduce a scholarly, interpretative approach to the teaching of science such as is now available in other departments of a university where the works of man are the subject of instruction. It is significant that this approach is now receiving the support of scientists. Joseph Fruton writes:

> There are many "gaps" between the man of art and the man of science, because of the impossibility of mastering the ideas, techniques, and language of more than a few compartments of human effort. The hope of bridging

these gaps through liberal education is, I believe, illusory. It may be suggested instead that it is the main function of liberal education to deepen understanding among non-artists or non-scientists beyond mere taste in some of the arts and mere utility in some of the sciences. The exercise of this function belongs to scholarship. According to this view, a scholar, though not a professional artist or a professional scientist, has acquired enough knowledge of the ideas, techniques, and the language of some arts or some sciences to permit him to examine critically the world of a group of artists or a group of scientists not only in terms of taste, utility, or some other social values, but also in terms of inspiration and craftsmanship. He does this by adding a comprehension of the craft to the skills and judgments of a historian, a biographer, or a philosopher.[11]

There are even scientists who, like Polycarp Kusch, will maintain that "from the point of view of the layman, it is more important to understand about science than to understand the content of science." [12] And the historical approach has found favor with Joseph Bronowski not only as an effective method of presenting the knowledge of science but as a direct approach to an understanding of certain aspects of the method of science:

> I think we need to teach science, even at school, not as a collection but as an evolution of knowledge. I think this important for three reasons. Because it sees science as a historical development, it opens links with history and literature and geography that can give help and a vivid perspective to the non-scientist. Because it presents science as changing, questioning, and argumentative, it can teach the methods of rational debate to everyone in the classroom, and this can be a life long lesson. But most important, the evolution of science goes to the heart of the scientific method: for it shows at each step how the logical deduction from what seems to lie behind the known facts must be confronted with experience.[13]

It would be dogmatic and even presumptuous at this stage to prescribe what course or courses must eventuate in conformity with a practicable new approach. A systematic historical survey of science may not be quite the answer, for the same reason that year-course surveys of the entire range of

English literature have never proved very effective—they are more likely to mean something to the student who knows a great deal than to one who knows little or nothing. Experience will have to determine in time what precise approach and what precise organization and form of presentation will best serve the end in view. Here and there efforts are already being made which will offer a basis for judging the effectiveness of particular methods. In the case of one particular type of course, considerable information exists. In 1946, in the Terry Lectures, published the next year as *On Understanding Science*, James Conant proposed a course in what he called the Tactics and Strategy of Science, to be taught by means of case histories. In 1949 he reported on the experience with such a course at Harvard College in a pamphlet, *The Growth of the Experimental Sciences. An Experiment in General Education.* A number of excellent case histories prepared for this course have also been published. Other approaches are possible.[14] Scientific knowledge has a way of growing out of a preoccupation of scientists with a particular type of phenomenon. It is possible to follow the sequence of significant experiments and observations, theories, and concepts which combined to provide a new idea of the behavior of nature and established what amounts often to a new science and technology. The studies of motion during the sixteenth and seventeenth centuries are a case in point. Electricity might be another—the interest in "static" electricity, the idea of a current, the relation of magnetism and electricity. Light is another—a particularly fascinating case, since it involved the rival theories of wave and corpuscular transmission, and the idea of ether, which played a curious role until experimental evidence ruled it out with impressive results. In biology, the concept of evolution might be a point of focus, or the steps by which the science of genetics was established. Though something of this approach occasionally gets into conventional introductory courses, in the new, the historical and philosophic aspects of these developments would be highlighted, as well as the methodological. One might, in a similar spirit, organize a course around the "classics" of scientific discovery as one now does with the classics of world literature. More than one type of course for non-scientists could be offered, one supplementing the other, or students might be encouraged to undertake one conventional college introductory course in a science of their choice before undertaking a course based on the new approach. In time the reasonable and useful will be found out and the fanciful and overambitious ruled out.

Such courses must be based on the assumption that the student is not going to be a scientist, but they cannot be based on complete ignorance. Acquaintance with more conventional courses in secondary schools would be taken for granted, and since instruments are essential to scientific investigation, some laboratory experience should be provided. High-school students are more likely to accept the content of a science and the technical rules for procedure without asking searching questions about subject matter and method, questions which become increasingly interesting with maturity. It is fortunately reasonable to expect that in the near future the scientific education of high-school students will be better than it has been in the past. Some of the country's leading scientists have assisted in the reformation of scientific textbooks and courses for secondary schools. At the college level there have been so far much less concern and much less effort on the part of scientists. The situation is ready for fundamental overhauling and serious experimentation.

Scientists will no doubt continue to be unhappy because the new approaches will provide even more opportunity to dodge the necessary acquaintance with mathematics. This is a touchy matter, and I have often sympathized with the impatience of scientists over the students who show an obtuse unwillingness to accept the discipline of mathematics. However, it is necessary to recognize that nature does not distribute her gifts alike, and that the lack of a taste for mathematics on the part of some can be paralleled by an equally obtuse lack of a taste for poetry on the part of others. If scientists can show indulgence toward one of their Ph.D. candidates who has trouble passing the reading examinations in foreign language for the doctorate, they might conceivably show a similar indulgence toward a student of languages who has trouble with mathematics. The important consideration is whether some improvement over the present situation is not somehow possible. Until now students have been expected to show capability in manipulating mathematical symbols and using them to solve problems in a science as a way of learning what mathematics in science is about. This may very well be like asking a student to write a fugue in order to have the music of Bach explained to him. The average non-scientist never really catches on to the nature of the mathematics he learns except blindly as a way of solving certain problems in conformity to certain arbitrary rules, and he learns even less why this sort of ingenious intellectual game turns out to be useful in science, even when he is able to solve the assigned problems. Though improvements in mathematical instruction are rapidly showing signs of resolving the first difficulty, there is little evidence that instruction in science is doing much with the second as a way of inculcating a more discerning notion about method. Introductory classes in physics are sometimes told that Newton found it necessary to invent the calculus; but if a student ever asked why this was necessary he was usually told that a course in calculus would provide the answer but that in his present ignorance explanation was hardly possible. The student's question can, however, be given a reasonably good answer if he has had algebra. He can be told what phenomena could not be readily expressed in the mathematics available to Newton, and what it is about calculus which makes it possible to bring such phenomena under control. It is not beyond the powers of a good instructor to make this quite clear, nor beyond the powers of an intelligent student to understand. In the same way, a student can be shown the difference between the whole tone scale and the diatonic scale, and by means of a few illustrations made to realize what musical possibilities are open to one and not to the other. This kind of understanding is, after all, more important for the non-scientist than learning to solve problems with an instrument the theory of which he does not fully understand and for the applications of which he has little aptitude.

The main difficulty in the way of establishing new courses for non-scientists will not be in the lack of agreement about the approach or about whether they are necessary or desirable; it will be largely a question of who is going to teach them. Knowing science is not the same as knowing about science, and in the present period of pressure to produce new scientists and to advance scientific knowledge, it is unlikely that bright and able scientists will want to divert their attention to any tasks other than those which they directly owe to science. Scientists who have already made their mark and who have a taste for philosophy and history might be induced to turn to the establishment of new courses as a way of getting involved in a new interest, and there has been for some years an increasing though still small number of historians of science who may have the capacity to work at the problem of organizing and trying out new courses along various lines. In the long run, however, we must look forward to the creation of a new type of academic figure, the scholar in science, who will have a relationship to the scientist analogous to that which the musicologist and music historian now bear to the composer and performer, or the professor of literature bears to the poet, the dramatist, and the novelist. These men may find an academic home in the already existing framework of departments of general studies or of the philosophy and history of science, where in time they may establish themselves as valuable and honored additions to the world of learning and education.

The most difficult question which remains serious and troublesome is the whole area of modern science, where the gap between the scientist and non-scientist is the greatest. There exists a feeling among scientists that the difficulties in making appreciable amounts of this recondite learning available to the layman may well be insurmountable. Conant stops short of the twentieth century in his course on the "Tactics and Strategy of Science," and on the most advanced modern theories in physics he has this to say:

> That matter disappears under certain circumstances and energy takes its place is not too difficult a conception to fit into a common sense framework. . . . But what has the speed of light to do with the whole business? That is the disturbing question; or rather, the answer is disturbing, for the scientist must say to the inquirer, "I'm sorry but that comes out of the theory of relativity, and it's

very difficult, if not impossible, to explain without a bit of physics and mathematics. I can't even give a decent hint about it; you'll have to take it on faith." [15]

It is possible that this view is too pessimistic. There is always an initial gap between avant garde learning and art and the outsider. Music, for example, that is at first incomprehensible becomes normal to the generation of youths who grow up listening to it. But it may be that the case with modern science is fundamentally different. If it is indeed true that on some matters of the greatest importance to a modern scientist the answer to the non-scientist must be that it is impossible even to give a hint of it to the uninitiated, then however hard we try, we have not made any real headway. The difficulty, moreover, is likely to become greater as science becomes increasingly refined and recondite, and if the best one will be able to do is to put off the questioner, like a kind father telling his child that he is not old enough—only in this case he never will be old enough—then the state of learning has become tragic. The very notion of asking an intelligent reader to take *science* on *faith* is a contradiction, a chilling paradox. If scientists now complain that the public image of them is naïve, then in the future they will have no protection against the combination magician, Santa Claus, and father image which the increasing separation of the scientist from the rest of us will induce.

The really important purpose in providing the proper scientific education for the non-scientist is not to make him understand how gadgets work or to prepare him for an enlightened voice in public decisions involving science—both greatly exaggerated and both of secondary importance. The real purpose—next to satisfying the normal curiosity of intelligent people—is to make some advance toward preserving the homogeneity of our culture. If science and its place in our culture were better understood than they are now by those who do not practice it, science would become a more familiar part of the common intellectual world which educated men share with one another. It is from the unfortunate plight of the non-scientist that this matter is usually considered, but the scientist has at least as much to gain, for science could then maintain a healthier relation to its society than at present. No improvement in this respect is possible, however, if the layman is painstakingly brought to the threshold of his own times and then told firmly, if with regret, that he cannot qualify for admission. If science continues to become increasingly essential to the perpetuation of our advanced form of civilization while at the same time it becomes increasingly incomprehensible except to scientists, then the inevitable consequence will be that the scientists will become isolated, separated from the rest of their society as a privileged elite—privileged intellectually because of the private character of their learning, and socially because of the enormous influence which they alone can exercise. This state of affairs would be lamentable for society generally, but also for the scientists as well; for they would then invite the fate which has usually overtaken such powerful, isolated elite groups in the past—they would first be looked upon with awe, then they would be feared, and finally they would be loathed. Bacon pictured such a scientific elite priesthood in the inhabitants of Salomon's House in the *New Atlantis*. Like all utopias, this one is founded on the assumption of a static order, frozen at the point of development which suits the writer's dreams and ideals, and accordingly it leaves out of consideration certain fundamental properties of man's nature. If the *New Atlantis* had been written as an imaginary history instead of a utopia, Bacon would have had to carry the story to its inevitable climax when the technological slaves of Bensalem finally rose up and did for Salomon's House what the lower orders did in France for the Bastille.

The disaster described in this philosophical myth represents an unlikely contingency in a distant mythical future, but like all myths this one adumbrates a reality. There is a threat, now widely recognized, to the well-being of our total culture in the increasing specialization and refinement of our learning in all fields. It is also recognized that this threat is represented in most acute form by science. A science becoming increasingly independent of the rest of the learning of our society, already taking up a large share of the national income and asking for more, revealing wonders that can be known only to the highest of its high priests, creating devices of unparalleled ingenuity for use and destruction at a rate progressively faster than society can adapt to them, freely taking over the air above and the sea and earth below without an assured knowledge of the possible consequences, containing no inherent imperative except the one to know, and propagating itself in ever larger forms—such a science could become monstrous. The increasing isolation of science is, moreover, becoming a form of exile. The scientists may be right in maintaining that the humanities have not been so well informed as they should be about science and particularly about modern science, but in the present advanced state of scientific knowledge, the assimilation of science into the whole of our culture cannot take place without the help of the scientists in promoting an understanding of the human relevance and meaning of their work. To do this they must become the allies of the humanities. The Director of the Oak Ridge National Laboratory recently remarked, "In making our choices we should remember the experience of other civilizations. Those cultures which have devoted too much of their talent to monuments which had nothing to do with the real issues of human well-being have fallen upon bad days." [16]

There is no practical alternative at the moment to acting on the assumption that it is not beyond our powers to prevent the most menacing aspects of our present advanced learning from becoming malign and destroying the organic unity of the body of our culture. In the teaching of science to undergraduates, there exists a modest but significant opportunity to take a step in the right direction. Such a step cannot be taken immediately without serious assistance from the present group of academic scientists. It is true they are very busy advancing scientific knowledge and training even more scientists to carry the scientific adventure even further, but the making of their knowledge available to others who will not be scientists may in the long run prove to be a responsibility of at least equal importance.

Notes

CHAPTER I

1. Conant, *Modern Science and Modern Man* (New York, 1959), pp. 145–46.
2. Riesman, "The Academic Career: Notes on Recruitment and Colleagueship," *Daedalus*, LXXXVIII (Winter, 1959), 156.
3. I do not have reference here to the notion of "the two cultures" which has been given prominence by C. P. Snow's book of that name—an important matter which requires separate consideration.
4. Bronowski, "The Educated Man in 1984," *Science*, CXXIII (1956), 710.
5. Nagel, "The Place of Science in a Liberal Education," *Daedalus*, LXXXVIII (Winter, 1959), 58.
6. Harcourt Brown (ed.), *Science and the Creative Spirit* (Toronto, 1958), p. xiii. (Published for the American Council of Learned Societies by the University of Toronto Press.)
7. Rabinowitch, "Science and the Humanities in Education," *Bulletin of the American Association of University Professors*, XLIV (1958), 458.
8. Jones, *One Great Society* (New York, 1959), p. 85.
9. For example, Lloyd Fallers, "C. P. Snow and the Third Culture," *Bulletin of the Atomic Scientists*, XVII (October, 1961), 306–10. For a more specialized notion of a bridge between the sciences and the humanities, see Derek Price, *Science since Babylon* (New Haven, 1961), Chap. VI. A slightly less inclusive summary than mine of the kinds of activities that might be termed "humanistic" is contained in *Science and the Creative Spirit*, p. xviii: "The humanities as we shall discuss them include: the creative arts, largely

interpreted; the various critical and historical disciplines that describe, explain, elaborate, or theorize on the basis of these arts; and the theory of such discussions, the relevant aspects of knowledge, of history, of criticism, aesthetic experience, imaginative creation."

10. R. S. Crane, "The Idea of the Humanities," *Carleton College Bulletin*, XLIX (1935), 9.

11. William S. Beck, *Modern Science and the Nature of Life* (New York, 1957), p. 34.

12. For criticism of the idea that the distinctive character of science is that it is cumulative and progressive, see Karl Deutsch, "Scientific and Humanistic Knowledge in the Growth of Civilization," in *Science and the Creative Spirit*, pp. 6–9; and David Hawkins, "The Creativity of Science," *ibid.*, pp. 131–33.

13. Conant, *op. cit.*, p. 97.

14. *Ibid.*, pp. 106–07.

15. Nagel, *op. cit.*, p. 57.

16. Bronowski, *Science and Human Values* (New York, 1956), p. 14.

17. *Ibid.*, p. 17.

18. Conant, *op. cit.*, p. 99.

19. *Science and Human Values*, pp. 65–66.

20. *Ibid.*, pp. 82–83.

21. M. E. Prior, "Bacon's Man of Science," *Journal of the History of Ideas*, XV (1954), 348–70.

22. For example, Conant, who finds the "justification of their [the scientists'] work in the pure joy of creativeness" (*Modern Science and Modern Man*, p. 99), and who would judge a scientific theory by its capacity to generate more scientific theories of a similarly fruitful nature: "This dynamic quality of science viewed not as a practical undertaking but as a development of conceptual schemes seems to me to be close to the heart of the best definition" (*On Understanding Science*, p. 24).

23. *Science and Human Values*, p. 77.

24. The mistake in the theory of the naturalistic novelists, in its most dogmatic form, was their notion that they were as much "outside" their materials as the physicist is "outside" the events he is studying in his laboratory. In their elaborate documentation, their attention to local details, their ruthless probing, their attitude of detachment, they introduced into the method of narrative literature a convention which had something in common with science, from which their attitude was borrowed. But the entire effect of their novels belies the presumptions of their method. They create sympathies, disgusts, at times a feeling of pathos—all of which implies not detachment but response to the human experience in terms of values and meanings.

25. See *Science and the Creative Spirit*, pp. xv–xvi, where the further point is made that "a work of art can be translated from one idiom to another only very approximately."

A stimulating discussion of the nature of science which came to my attention too late to have its views incorporated in this essay in definition is that of Stephen Toulmin, *Foresight and Understanding* (Bloomington, 1961). An important feature of this book is the criticism of most approaches to definition and of the importance usually attached to prediction as an essential element of science. I do not believe that Toulmin's argument affects my choice of those aspects of science which most effectively distinguish it from the humanities.

CHAPTER II

1. It is a widely accepted commonplace of popular scientific history that the most striking consequence of the Copernican theory was that, by placing the sun rather than the earth at the center of the planetary system, it destroyed the comforting illusion that we occupied a central and therefore favored place in the universe and thus plunged mankind into gloom. I myself have not come across any contemporary expressions of precisely this notion. It may well be a later deduction of what ought to have been the popular reaction to the Copernican system.

2. Boyle, *Works* (1772) II, 20.

3. Russell, *The Impact of Science on Society* (New York, 1953), p. 14.

4. Russell, "My Philosophical Development," *Encounter*, XII (February, 1959), 24.

5. This aspect of the modern situation is considered by Erich Heller in *The Disinherited Mind* (New York, 1952), throughout but particularly in the essay on Goethe.

6. For a precise statement of this view of the social sciences, see Henry W. Rieken, "The Social Sciences and the Federal Government," *The Midwest Sociologist*, XXI (1959), 66.

7. Bronowski, *Science and Human Values* (New York, 1956), pp. 48–49.

8. *Ibid.*, pp. 87–88. For an earlier form of this argument see Karl Pearson, "Science and Citizenship," in *The Grammar of Science* (New York, 1937), pp. 11–12.

9. Ernest Nagel, "The Place of Science in a Liberal Education," *Daedalus*, LXXXVIII (1959), 62.

10. Bronowski, *op. cit.*, p. 77.

11. Snow, *Science and Government* (London, 1961), p. 76: "Again unfortunately, the constraints of secrecy, though they disturb the comparative judgment, do not disturb the scientific process. In more liberal days, in the days of Rutherford's Cambridge, Bohr's Copenhagen, Franck's Göttingen, scientists tended to assume, as an optimistic act of faith, as something which ought to be true be-

cause it made life sweeter, that science could only flourish in the free air. I wish it were so. I think everyone who has ever witnessed secret science and secret choices wishes it were so. But nearly all the evidence is dead against it. Science needs discussion, yes: it needs the criticism of other scientists: but that can be made to exist, and of course has been made to exist, in the most secret projects. Scientists have worked, apparently happily, and certainly effectively, in conditions which would have been thought the negation of science by the great free-minded practitioners. But the secret, the closed, the climate which to earlier scientists would have been morally intolerable, soon becomes easy to tolerate. I even doubt whether, if one could compare the rate of advance in one of the secret sciences with one of those which is still open to the world, there would be any significant difference. It is a pity."

12. Conant, *On Understanding Science* (New Haven, 1947), p. 10.

CHAPTER III

1. Jones, *One Great Society* (New York, 1959), p. 3.
2. Conant, *On Understanding Science* (New Haven, 1947), p. 2.
3. Bronowski, "The Educated Man in 1984," *Science*, CXXIII (1956), 710.
4. Gallant, "Literature, Science, and the Manpower Crisis," *Science*, CXXV (1957), 788.
5. Snow, *The Two Cultures and the Scientific Revolution* (New York, 1959), pp. 4–6.
6. *Ibid.*, p. 4.
7. *Ibid.*, pp. 4–5.
8. The following supporting statement is from *The Complete Scientist*, the Report of the Leverhulme Study Group to the British Association for the Advancement of Science (London, 1961), p. xvi: "As the frontiers of knowledge are extended, workers on the perimeter are in danger of finding themselves isolated in their outposts by ever-increasing distances, not only from one another but also from the main body of those who, because of limitations in their education, have no understanding of or interest in the direction or significance of these advances."
9. Oppenheimer, "Tradition and Discovery," *ACLS Newsletter*, X (October, 1959), 12–13.
10. *Ibid.*, p. 13.
11. *The Two Cultures*, pp. 17–18.
12. Bronowski, *Science and Human Values* (New York, 1956), p. 55.
13. Gallant, *op. cit.*, p. 788; also p. 790.
14. *Ibid.*, p. 788.
15. Rabinowitch, "Science and the Humanities in Education," *Bulletin*

of the American Association of University Professors (hereinafter
cited as *AAUP Bulletin*), XLIV (1958), 457.

16. *The Two Cultures*, p. 6.
17. *Ibid.*, p. 34.
18. *Ibid.*, p. 14. The interest of scientists in music is frequently com-
    mented upon, most often, however, as a convincing way of demon-
    strating that scientists are cultivated men. For example: "In our
    time, it seems likely that the scientists are on the whole more musi-
    cal than men of letters. This may be our best answer, if we want
    one, to the charge that scientists are not barbarians." (Erwin
    Klingsberg in *Bulletin of the Atomic Scientists*, XV [1959], 401);
    "There need be no dichotomy [between science and culture]. It
    was not inconsistent for Einstein to love the violin" (*Science*,
    CXXV [1957], 790).
19. Conant, *Modern Science and Modern Man*, p. 147.
20. Harcourt Brown, in *Science and the Creative Spirit* (Toronto,
    1958), pp. xxi–xxii.
21. Auden, "The Alienated City," *Encounter*, XVII (1961), 14.
22. René Dubos, *The Dreams of Reason; Science and Utopias* (New
    York, 1961), p. 62.
23. Derek Price, *Science since Babylon* (New Haven, 1961), p. 117.
24. Erich Heller, "Faust's Damnation. The Morality of Knowledge,"
    in *The Listener* (January 11, 1962), p. 60. This is the first of three
    BBC talks printed in *The Listener*.
25. Glass, "The Academic Scientist: 1940–1960," *AAUP Bulletin*,
    XLVI (1960), 155.
26. Glass, *Science and Liberal Education* (Baton Rouge, 1959), p. 85.
27. *On Understanding Science*, pp. 21–22.
28. *Science*, CXXXII (December, 1960), 1803–04.

CHAPTER IV

1. Gallant, "Literature, Science, and the Manpower Crisis," *Science*,
   CXXV (1957), 788.
2. National Science Foundation, *Reviews of Data on Research and
   Development* (Sept., 1961), and Bentley Glass, "The Academic
   Scientist: 1940–1960," *AAUP Bulletin*, XLVI (1960), 151–52.
   Compare the estimate of expenditures given by Don Price as of the
   year 1953: "A half century ago the annual federal expenditures for
   research and development were in the range of ten million dollars.
   By 1930 they were something less—perhaps considerably less—than
   a hundred million dollars. They reached a billion dollars by the
   end of World War II, and two billion about a year ago" (*Government and Science* [1954], p. 35). On the matter of future spending
   and the choices to be made, see Alvin M. Weinberg, "Impact of

Large-Scale Science on the United States," *Science*, CXXXIV (1961), 161–64.

3. An excellent general review of the problem of science and government, its history in America, and its present implications, is that of Don K. Price, *Government and Science* (New York, 1954).
4. Snow, *Science and Government* (London, 1961), pp. 55–56.
5. Conant, *On Understanding Science* (New Haven, 1947), p. 4.
6. Don Price, *op. cit.*, p. 169.
7. Rabinowitch, "Science and Humanities in Education," *AAUP Bulletin*, XLIV (1958), 454.
8. *Science and Government*, pp. 1–53.
9. *On Understanding Science*, p. 10.
10. Serious doubts have been cast on Snow's interpretation of the events involving these two men, and until more extensive historical accounts have appeared one has to suspend judgment. However, since Snow bases his recommendations on what he regards as the facts in the story, the reader of *Science and Government* should consider the views of those who also had some first hand knowledge of the men and the circumstances and who read the record differently from Snow. The two following items, though a small sampling from among the many reviews of Snow's book, will serve this purpose: A. J. P. Taylor, "Lindemann and Tizard: More Luck than Judgment?" *The Observer* (London, 9 April 1961), and the anonymous review in *The Times Literary Supplement* (London, 14 April 1961), p. 226.
11. *Science and Government*, pp. 80–81.
12. *Ibid.*, pp. 82–83.
13. *Ibid.*, p. 82.
14. Snow, *The Two Cultures and the Scientific Revolution* (New York, 1959), p. 50. See also "Afterthoughts," *Encounter*, XIV (1960), 65.
15. Since impartiality, a valuable quality in an administrator, is one of the virtues most often attributed to scientific training, the following remarks of Conant are especially relevant: "Therefore, to put the scientist on a pedestal because he is an impartial inquirer is to misunderstand the situation entirely. Rather, if we seek to spread more widely among the population the desire to seek the facts without prejudice, we should pick our modern examples from the non-scientific fields. We should examine and admire the conduct of the relatively few who in the midst of human affairs can courageously, honestly, and intelligently come to conclusions based on reason without regard for their own or other people's loyalties and interests, and having come to their conclusions, can state them fairly, stick by them, and act accordingly" (*On Understanding Science*, pp. 9–10).
16. *Government and Science*, p. 203.
17. *Ibid.*, pp. 202–03.

CHAPTER V

1. Glass, "The Academic Scientist: 1940–1960," AAUP *Bulletin,* XLVI (1960), 155.
2. Gallant, "Literature, Science, and the Manpower Crisis," *Science,* CXXV (1957), 789.
3. There exists also a curious unofficial pecking order within the sciences. My own entirely informal observations lead me to believe that generally speaking physicists think of themselves at the top of the order, and certainly above the chemists, and so on. There are also gradations within individual sciences: theoretical physicists seem to consider themselves above the experimental, for instance. In general, all scientists tend to regard themselves as engaged in more basic and hence superior work than engineers. I have sensed also a feeling of gradation within the fraternity of engineers. Mathematicians, of course, tend to regard themselves as the most Olympian of all. The reasons for this tacit hierarchy might prove of interest to sociologists.
4. Conant, *On Understanding Science* (New Haven, 1947), pp. 7–8, 9.
5. Gallant, *op. cit.,* p. 790.
6. David Hawkins, "The Creativity of Science," in *Science and the Creative Spirit* (Toronto, 1958), pp. 130–31.
7. "Scientists and Laymen," *The Key Reporter,* XXVI (1961), 2–3.
8. Gallant, *op. cit.,* p. 790.
9. Bronowski, *Science and Human Values* (New York, 1956), pp. 90–91.
10. Conant, *The Growth of the Experimental Sciences. An Experiment in General Education* (Cambridge, 1949), pp. 10–11.
11. Fruton, "The Arts, the Sciences, and Scholarship," *The Yale Review,* L (Winter, 1960), 224. This is an excellent and important article. My own views parallel the argument at many points.
12. Kusch, *op. cit.,* p. 4.
13. Bronowski, "The Educated Man in 1984," *Science,* CXXIII (1956), 712.
14. Several already in use are referred to by Conant in the pamphlet cited.
15. Conant, *Modern Science and Modern Man* (New York, 1959), p. 64.
16. Alvin M. Weinberg, "Impact of Large-Scale Science on the United States," *Science,* CXXXIV (1961), 164.